Energy Resilient Buildings & Communities:
A Practical Guide

Energy Resilient Buildings & Communities:
A Practical Guide

by
Brian Levite and Alexander Rakow

THE FAIRMONT PRESS, INC.

CRC Press
Taylor & Francis Group

Library of Congress Cataloging-in-Publication Data

Levite, Brian.
 Energy resilient buildings & communities : a practical guide / by Brian Levite and Alexander Rakow.
    pages cm
 Includes bibliographical references and index.
    ISBN 0-88173-718-6 (alk. paper) -- ISBN 978-1-4987-2959-8 (Taylor & Francis distribution : alk. paper) -- ISBN 0-88173-719-4 (electronic) 1. Energy conservation--Handbooks, manuals, etc. I. Rakow, Alexander. II. Title.

TJ163.3.L48 2015
690.028'6--dc23

2014047847

Energy Resilient Buildings & Communities: A Practical Guide/Brian Levite, Alexander Rakow.
©2015 by Hitachi Consulting Corporation. All rights reserved. No part of this publication may be reproduced or transmitted in any form or by any means, electronic or mechanical, including photocopy, recording, or any information storage and retrieval system, without permission in writing from the publisher.

Published by The Fairmont Press, Inc.
700 Indian Trail
Lilburn, GA 30047
tel: 770-925-9388; fax: 770-381-9865
http://www.fairmontpress.com

Distributed by Taylor & Francis Ltd.
6000 Broken Sound Parkway NW, Suite 300
Boca Raton, FL 33487, USA
E-mail: orders@crcpress.com

Distributed by Taylor & Francis Ltd.
23-25 Blades Court
Deodar Road
London SW15 2NU, UK
E-mail: uk.tandf@thomsonpublishingservices.co.uk

Printed in the United States of America
10 9 8 7 6 5 4 3 2 1

ISBN 0-88173-718-6 (The Fairmont Press, Inc.)
ISBN 978-1-4987-2959-8 (Taylor & Francis Ltd.)

While every effort is made to provide dependable information, the publisher, authors, and editors cannot be held responsible for any errors or omissions.

The views expressed herein do not necessarily reflect those of the publisher.

Table of Contents

Introduction ix

Part I—Energy Resilience Imperatives and Local Advantages. 1
Chapter 1—Today's Energy Challenges 3
    Supply and Demand in the US Energy System 4
    Water and the Energy System 8
    Energy Infrastructure 11
    Climate Change 15
    System Vulnerabilities to Sabotage 21

Chapter 2—Local Energy Resilience 27
    Energy Resilience vs. Energy Independence 27
    The Importance of Local Energy Resilience 28
    Finding a Place on the Energy Resilience Spectrum 30
    Energy Resilience in Our Four Community Types 36

Part II—Energy Resilience Management 51
Chapter 3—Institutional Planning for Energy Resilience 53
    Identify/Convene Stakeholders 60
    Form Leadership Team 66
    Develop An Energy Vision 67
    Develop An Energy Resilience Baseline 67
    Develop Specific Goals 67
    Evaluate and Rank Programs 69
    Explore Funding Sources 73
    Compile the Plan 78
    Measurement, Verification and Plan Alterations 79

Chapter 4—Energy Resilience Maturity Model 85
    Goals 87
    Strategy 88
    Project Financing 90
    People and Maintenance 94
    Efficiency Technology 94

    Energy Generation... 94
    Emergency Preparedness... 97
    Purchasing... 98
    Information Management... 100

Part III—Energy Resilience Performance Areas ... 103
Chapter 5—Energy Efficiency Approaches ... 105
    Energy Efficiency and Conservation... 106
    Tailoring Demand Reduction to Meet Resilience Goals ... 107
    Continuous Energy Performance Improvement ... 108
    Practical Solutions for Energy Planning... 125

Chapter 6—Community Energy Generation ... 131
    Drives for Distributed Generation... 132
    Evaluating Your Power Needs ... 135
    Technology Overviews ... 150
    Microgrids... 170

Chapter 7—Planning for an Energy Emergency ... 179
    Risk Assessment... 180
    The Four Phases of Emergency Management... 186
    Mitigation ... 187
    Preparation... 188
    Response... 190
    Recovery... 191
    Conclusion... 195

Conclusion... 197

Citations ... 201

Index ... 213

Acknowledgements

This book is dedicated to the two little flowers born during its writing, Lilliana Levite and Magnolia Rakow. May they be both energetic and resilient in a changing world.

We would like to first thank our families for their support during the research and writing of this book—particularly our wives who did more than their share while we worked on this second full-time job.

We would like to thank our spectacular editors and technical reviewers, Christine Meyer-Oertel of Avenue 3 Communications and Stacey Toevs, Coleman Adams and Michael Uhl of Hitachi Consulting.

We would like to thank Alexander Dane of the National Renewable Energy Laboratory for his technical advice, and valuable assistance in framing this project.

Finally, we would like to thank the entire team at Hitachi Consulting, particularly our team leader, Todd Price, for the support and encouragement we received along the way.

Introduction

In October 2012, Hurricane Sandy battered the densely populated coastal areas of the eastern United States, leaving more than 8.5 million people without the power they rely on to light and heat their homes, run their businesses, and perform the essential functions of government. A week after the storm, 1.3 million of these people were still in the dark. Due to the fragility and interconnectedness of our power grid, the victims of the storm were not limited to the east coast. The storm caused outages as far inland as Indiana and Michigan. Hurricane Sandy was a particularly destructive storm, but it was by no means the only such event to cause widespread blackouts in recent years. The number of power outages from storms has been increasing steadily in the United States since the 1980s, and has doubled since 2003.[1]

Severe weather represents only one of many serious threats that face the American energy system. Water scarcity, aging infrastructure, terrorism, and even price instability all threaten to disrupt the steady flow of energy on which our society relies. Each of these threats is growing and, every day, the task of protecting our energy security grows along with them. In this book, we examine the energy security imperative from the perspective of an individual community, and describe the ways in which these communities can become more resilient to shocks to our energy system.

LOCAL ENERGY RESILIENCE OVER NATIONAL ENERGY INDEPENDENCE

Energy security is a concept most often discussed on a national scale. Lessening our dependence on foreign oil is a perennial campaign theme in national politics out of concern over geopolitical security, and the security of our energy supply. However, the ideal of energy independence has proven elusive, and in recent years, the value of absolute energy independence as a goal has come into question. Increasingly, the national conversation around energy security has shifted from a debate focused on independence from foreign sources of oil to a more nuanced discussion of what's called 'energy resilience.'

Instead of seeking energy independence, we should instead seek strategies to weather shocks to our energy system, whether they be economic (price spikes), physical (weakened energy infrastructure), or climatological (severe storms, water shortages, etc.). *We define energy resilience as the relative ability of an institution to carry out its mission during a shock to the energy system.* This shock may be a disruption of foreign fuel supplies, but it might also be the result of severe storm, a water shortage, an equipment failure, or an act of sabotage. Unlike energy independence, which is often described as a binary state, we use the term energy resilience to describe a spectrum. Each institution occupies a particular spot on this spectrum based on their readiness to continue operations during an energy disruption, and each may identify a different spot on the spectrum that they would like to ultimately achieve.

Even as energy resilience has become a popular paradigm for approaching energy security, it has largely retained a national or regional unit of analysis. Most strategies and policy prescriptions proposed have been geared toward regional security—focusing on grid-level upgrades and state or multi-state preparedness. Following Hurricane Sandy, both Mayor Bloomberg of New York and Governor Christie in New Jersey rolled out disaster resilience plans. These plans included actions to harden infrastructure and strengthen emergency response resources. To some degree, these measures will mitigate the effects of the next great storm or other regional disaster on the energy security of individual institutions within that region. However, administrators in charge of those institutions may not want to leave the protection of their energy supply, and the vital missions that it supports, to regional plans and initiatives.

In this book, we argue that it is more pressing, and indeed more fruitful, to approach energy resilience on a much smaller scale, at the level of a single site occupied by a single community or institution. A shock to the energy system affects a local institution differently than it does a whole a region. The good news is that local institutions have natural advantages that make high levels of energy resilience more attainable than they would be for an entire city or state, and the development of new technology and best practices make local energy resilience not just possible, but cost effective.

THE ENERGY RESILIENT COMMUNITY

For many small municipalities and local institutions, an on-site energy resilience plan is not only more practical than waiting for nationwide action, it's vital to carrying out their mission. A hospital, for example, cannot afford to put patient care or medical research in jeopardy waiting on national political action to combat increasing energy instability. As energy bills rise and severe weather events become more frequent, many hospital administrators are pursuing energy resilience as a way to protect their patients, and control the rising cost of healthcare. Similar imperatives exist to protect local functions in nearly every industry, and especially those that provide residential services.

The unit of analysis we use in this book is a single community, and the management prescriptions in the latter chapters of the book are directed at the managers and stakeholders of these communities. As we define the term, a community is not simply a neighborhood, or just any group like-minded individuals, but a very specific kind of entity. We define a community as an *institution occupying a single contiguous space, and with the ability to exercise some degree of central control over the facilities and activities that occupy that space*. This definition was chosen because it describes the type of institution where energy resilience is not just particularly important, but also more practical.

In this book, we draw on examples and lessons from four key community types (military bases, healthcare campuses, educational campuses and municipal governments), and describe a framework for developing an energy resilience plan that applies to each. These communities each have a vital mission that relies on a constant supply of power. Whether protecting national security and the safety of troops, treating patients, ensuring the safety of student residents, or providing emergency response services, none of the communities we examine can endure an extended power outage without suffering serious harm. These communities also share qualities that help them address this imperative. The space they occupy gives them advantages in developing on site generation and improving energy efficiency. The permanence of the institution that occupies the space, and the control they exert over the activities therein, gives it the ability to engage in large scale, long-term energy resilience planning. Although, we give special attention to these four community types throughout this book, our definition of community could include many other types of institutions, from corporate campuses to sports facilities and even shopping malls.

The focus of this book is also clearly on the United States. This choice was made because it is the market that we, the authors, have the most experience in and there are specific challenges to be met in the U.S. Other countries may have very similar challenges with aging infrastructure or severe weather or price instability, but the nature and the balance of those threats will be slightly different and other threats may be present. We wanted to be clear that our analysis does not specifically include the unique challenges of nations outside the U.S. That said, we believe that understanding energy resilience threats and conducting long-range energy resilience planning will benefit communities all over the globe. In Europe, the drivers of energy resilience may be the combination of high energy prices and unstable fuel supply from other nations. In Japan it may be combination a high percentage of energy imports combined with seismic instability. In the developing world, building resilient communities with local generation resources may simply be more cost effective than attempting to replicate the centralized model of first world countries. We sincerely hope this book will be valuable to a global audience, despite its research focus on the U.S.

THE ORGANIZATION OF THIS BOOK

This book is divided into three main parts, as described below:

Part One:
Energy Resilience—Imperatives and Local Advantages

Part one of the book is intended to give context to the energy resilience discussion. In Chapter 1, we describe the specific energy security threats that are facing local institutions and communities. We present the evidence for each, and describe how it could impact energy security at a community fitting our definition. In Chapter 2, we present the concept of energy resilience. We discuss the importance of the resilience framework to energy security, and describe why we feel it is particularly important at the level of a local community or institution. We describe how an energy shock can affect the mission at each of our four community types, and the advantages that each enjoy in their pursuit of energy resilience.

Part Two: Energy Resilience Management

The second part of the book shifts away from the background information presented in Part I, and toward more concrete guidance for pursuing energy resilience at a particular institution. Chapter 3 lays out a framework for how to pursue energy resilience from a management perspective. It discusses the steps involved in bringing together the relevant parties, identifying goals, and pursuing those goals in a manner that is most likely to achieve success. Chapter 4 presents a "maturity model" designed to allow managers to assess where their institution lies on the energy resilience spectrum, and plot a course toward where they would like to be. Unlike Leadership in Energy and Environmental Design (LEED) for green building or ENERGY STAR® for energy efficient buildings, there is no benchmarking system for energy resilient facilities or communities. Chapter Four presents the outline of a benchmarking system for the main performance areas of energy resilience.

Part Three: Energy Resilience Performance Areas
The final section of the book is dedicated to describing the three main tent poles of energy resilience performance: energy efficiency, on-site generation, and emergency planning. Chapter 5 lays out a process for maximizing energy efficiency at an institution, from auditing to the selection and execution of energy conservation measures. Chapter 6 does the same for power generation, describing the technologies an institution is likely use when developing an on-site generation portfolio or a microgrid. Finally, Chapter 7 provides guidance on the process of planning for an energy emergency, and ensuring that all of an institution's energy resources (including staff) are used to maximum effect in mitigating the damage of an energy disruption.

CASE STUDIES

Throughout the book we have included case studies of institutions in our four community types—military bases, higher education campuses, healthcare campuses, and local governments. These case studies are based on interviews we conducted with institutional administrators and managers about what drives their pursuit of energy resilience and what initiatives they've pursued to advance that goal. These case studies are included to provide real world examples of the challenges that institutional leaders are facing in terms of energy security, and the creative solutions they have deployed to advance energy resilience.

Realistically speaking, every institution has some room for improvement in energy resilience planning, even those described in these case studies. There is no community that is completely immune to threats to the energy system. However, one of the best ways for an institution to move further along the energy resilience spectrum is to learn from the successful strategies already employed by similar communities. The case studies in this book are intended to provide an opportunity for readers to see how other communities have attacked this problem and evaluate which strategies may be applicable to their institution.

HOW TO USE THIS BOOK

This book is written as a practical guide for those interested in the pursuit of energy resilience at a local scale. Anyone interested in energy security should find the content useful, but the book is especially tailored for a few key audiences. Primarily, the book is intended for institutional leaders interested in providing uninterrupted services to their constituents and planning for security in an increasingly insecure energy landscape. We selected case studies from institutions where a constant supply of energy is non-negotiable, and where administrators have a tremendous responsibility to ensure that the mission of their organization continues without interruption.

As we will discuss, however, energy resilience efforts can also help to reduce energy costs and minimize the environmental impact of our energy use. Facility managers tasked with saving their institution money by reducing consumption, demand charges, and exposure to energy risk will find a myriad of strategies for their pursuit of those goals. Professionals focused on environmental sustainability will also find strategies to support their efforts—whether those are focused on energy efficiency to reduce carbon emissions or the increased use of renewable energy. Ultimately, energy resilience must support the overarching mission of the institution. Whether certain actions save money or make an institution more environmentally sustainable, the North Star for institutional energy resilience is the preservation of business as usual inside a community during those times when chaos reigns outside its walls.

While we recommend reading this book from start to finish, the chapters are self-contained enough to be valuable reading material if you simply want to learn about one subject, such as threats to the energy system or how to develop a comprehensive energy efficiency plan. We have attempted to frame this information in as practical and actionable a way as possible. Our goal in writing the book was to advance the interest and expertise of the reader in energy resilience, encouraging more action on this front.

Part I

Energy Resilience—Imperatives and Local Advantages

Chapter 1

Today's Energy Challenges

The reliable access to energy we enjoy in the United States is based on a complex relationship of natural, technological and institutional factors. Each flip of the light switch relies on technology and institutions adequate to extract fuel from the natural environment, transform it into electricity, and distribute it over tens or even hundreds of miles. Each step in this process in turn relies upon social organization and support, from international markets for fossil fuels to the public/private coordination that allows for the generation and distribution of energy on a large scale. For such a complex process, it is remarkably reliable. However, when just one of the necessary natural, technological, or institutional conditions fails, we quickly find ourselves in the dark.

Based on the community-oriented perspective of this book, we discuss energy resilience as the ability of institutional managers and administrators to ensure a constant supply of energy adequate to guarantee the safety of their constituents and the uninterrupted pursuit of their institution's core mission. Energy resilience has different practical implications at our four community types (hospitals, military bases, educational campuses, and municipalities), but leaders in each should be aware of the factors that pose potential threats to their energy security, and be able to assess each threat to ensure that it is properly addressed. The most likely threats to energy resilience are price volatility, power quality, and full grid outages due to aging energy infrastructure, severe weather events, and threats from sabotage.

This chapter focuses on the nature and severity of threats to the natural, social, and technological systems on which the U.S. energy security relies. First, we describe the factors that affect overall supply and demand trends in our energy system. We will evaluate the potential for energy demand to outstrip the natural resources needed to supply it, and any effects this may have on price volatility. We will then look at the state of the technology necessary to deliver energy to individual communities, with a focus on distribution systems and energy grids. We will describe the threat to these technological systems from the natural forces of climate change and extreme weather.. Finally, we will analyze more active social factors that may affect energy supply, such as such as terrorism and social protest.

SUPPLY AND DEMAND IN THE US ENERGY SYSTEM

For decades, we have been struggling to make the transition from a world in which the supply of fossil fuels seemed endless, to one in which the limits on that supply are making the extraction of these resources increasingly more expensive and difficult. As we have approached the limits of existing extraction capacity, we have sought new, more expensive or technologically complex methods for gathering fuel. To find untapped reserves of fossil fuel, we have moved deeper into the ocean and further into the Arctic than previously thought possible. We have drilled in formerly protected lands, and devised technology to extract gas from shale formations that were previously unreachable. In doing so, we have managed to keep up with ever-rising demand. However, our own ingenuity will only hold out so long against a finite resource, and as the gap between effective supply and demand for fossil fuel widens in the future, new stresses will arise on our energy security, from price volatility to social conflict and climate change.

The immediate future of our fuel supply seems to be relatively secure. The U.S. Energy Information Administration (EIA) has reported that it expects the global supply of crude oil and other liquid hydrocarbons to be adequate to meet the world's demand for at least the next 25 years.[2] In the United States, the preliminary EIA 2014 Annual Energy Outlook describes a future of increased

Today's Energy Challenges

domestic fuel production and decreased per capita demand for energy.[3] The report predicts that the rate of domestic oil production will soon match its all-time high, set in 1970, of 9.6 million barrels a day—due largely to technological advancements that allow previously unreachable deposits to be effectively exploited.

The outlook is even brighter for domestic natural gas. The report predicts that, while oil production will begin to plateau around 2020, the production of natural gas will continue to grow into the foreseeable future. By the farthest time horizon for EIA predictions in 2040, natural gas production in the United States will have grown by 56 percent over 2012 levels.[4]

This growth rate will more than keep up with the growth in energy demand domestically, and will result in an overall decrease in the share of U.S. fossil fuel use from imports to 32 percent in 2040, down from its peak at over 60 percent in 2005. The report predicts that the United States will become an overall net exporter of natural gas as soon as 2018.

Two of the most important factors behind the country's ability to meet domestic energy demand are population growth and ener-

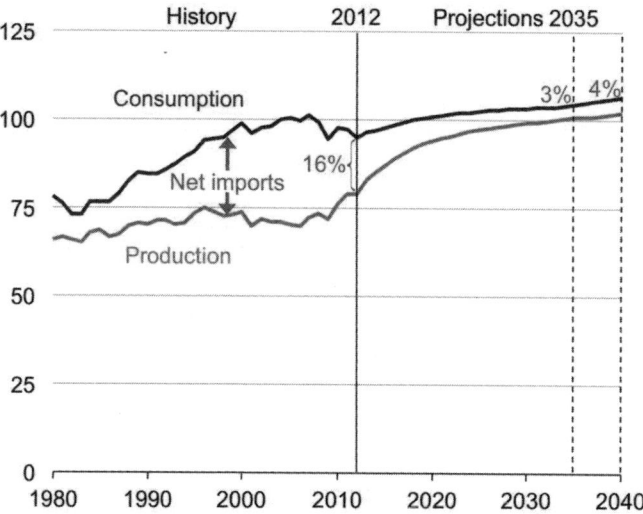

Figure 1-1: Total energy production and consumption, 1980-2040, quadrillion Btu. Source: U.S. EIA Annual Energy Outlook, 2014.[5]

gy intensity. The first is a measure of birth and immigration rates, and the second is a measure of how much energy each person uses, or how much energy is used per unit of economic activity. When the EIA analyzed these variables in 2013, they found that U.S. population is likely to increase by 0.9 percent per year from 2011 to 2040, and the economy, as measured by GDP, is likely to grow over that same period at an average annual rate of 2.5 percent. Meanwhile, the total energy consumption is expected to increase by 0.3 percent per year, which implies that energy intensity, both per capita and per dollar of GDP, will decline into the foreseeable future.[4] Much of this decrease in energy intensity is due to improvements in the efficiency of energy consuming equipment, used in buildings, appliances, cars and factories. Chapter 5 will discuss the ways that energy efficient technologies can be harnessed at the institutional level to improve energy resilience.

It might seem from these numbers that there is little cause for concern when it comes to the future of our energy supply. Indeed, if the EIA projections are borne out over the next 25 years,

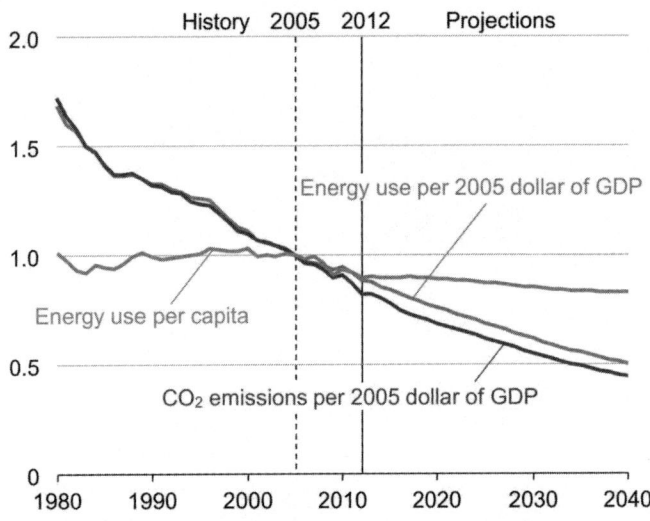

Figure 1-2: Energy use per capita, energy use per dollar of GDP, and emissions per dollar of GDP 1980-2040 (index: 2005 = 1). Source: US EIA Annual Energy Outlook, 2014.[4]

there should be ample global supply of fuel to meet energy demands in the United States, an increasing share of which would be produced domestically. However, a secure supply does not necessarily mean low prices, and the steadily rising cost of energy is compelling some communities to trim their loads, and become more energy resilient. The case study that follows describes a healthcare organization that was spurred toward energy efficiency by rising costs, but ended up fully embracing energy resilience.

> **Case Study—University Medical Center of Princeton at Plainsboro[45]:**
>
> Like hospital administrators all over the country, the leadership of the Princeton HealthCare System faced a growing dilemma at the beginning of this century. Healthcare costs were rising alongside the cost of doing business, suggesting an undesirable tradeoff between breadth of service and health of the business. As Princeton HealthCare System President and CEO Barry Rabner explained it, "in order to keep the costs for patients from rising, I needed to figure out where to cut back. When compared to cutting staff, or clinical services, or support for patients, cutting energy costs was a no brainer. Honestly, I look at the opportunity to save money on energy as a gift to me as an administrator."
>
> In recognition of this opportunity, Princeton HealthCare System has sought to make the newly opened University Medical Center of Princeton at Plainsboro (UMCPP) an example for the industry. Princeton worked with the energy contractor NRG Energy to integrate the suite of energy efficiency and generation technologies on display at UMCPP into one hyper efficient, integrated on-site energy system. A 250 kilowatt solar array works in conjunction with a massive $34 million CHP plant, which combined with a chilled water thermal energy storage system maintains the temperature of the buildings, all of which is continually monitored and controlled by a computer system which ensures optimal operational and fiscal efficiency.
>
> The upshot of this system is 100% power redundancy with the grid, with essential clinical equipment backed up with batteries.

> That means that UMCPP meets all of its energy needs with on-site generation, and uses the grid as a back-up option should its own systems fail. Should there be a simultaneous failure of both the on-site system and the grid, critical equipment is backed up by battery, or in limited cases, by small diesel generators. This arrangement sets a very high standard for energy resilience within the health care industry, and helps Princeton HealthCare System to advance other areas of its mission as well. As Rabnor explained, "we found a system that not only reduces our energy costs, but gives us security in the face of uncertainty in the energy market, which helps us protect our patients. It was that desire to protect our patients that ended up being our main driver in our decision to invest in this system."
>
> Although patients may not be aware of how much more secure the operations of UMCPP are when compared to a grid dependent hospital, steps that Princeton has taken to improve energy efficiency and environmental stewardship have had a positive effect on the patient experience as well. For instance, the building is designed to take maximum advantage of natural light. It is south-facing, and during the day 90% of interior spaces can be lit with sunlight, a rarity in hospital settings. In developing UMCPP, Princeton demolished an adjacent factory site, remediated the 32-acre plot on which it sat, and converted into a park, planted with indigenous trees and plants.

Although our supply of fuel may not be in danger of drying up in the near future, it is important to remember that fuel is not the only input in a robust energy system, and threats to water supplies and energy delivery infrastructure could soon affect energy security for local communities.

WATER AND THE ENERGY SYSTEM

Supply of fuel is just one variable in the formula that guarantees energy to our homes and business. Increasingly, scientists are singling out water as a possible limiting factor in this equation.[5]

Water is a key ingredient throughout the energy supply chain. It is used extensively in the drilling and mining of fuels, including natural gas, coal, oil, and uranium. Fuel refining and processing is a water-intensive process in its own right, and is required before oil, uranium, or natural gas can be used in a power plant. Transporting the refined fuel uses still more water. Water is used to test fuel pipelines for leaks, and to transport coal through slurries. Finally, water is an essential element in the vital functions of a vast majority of the power plants in the United States.[6]

About 90 percent of power generation in the United States is thermoelectric—a term used to describe the process of using heat to create steam and turn a turbine that generates electricity.[7] Thermoelectric power plants generate heat through the combustion of fossil fuel, the fission of nuclear fuel, or the concentration of solar radiation. Water is used to generate the steam that turns the turbines in these plants, and sometimes to cool steam to condense it for reuse. More freshwater is withdrawn from natural reserves for thermoelectric power generation than for any other purpose: over 40 percent of the total withdrawn in 2005, the last time a thorough analysis was performed.[8] In addition to withdrawing more freshwater than any other industry, energy generation is also one of the largest non-agricultural *consumers* of freshwater. Water consumption refers to the portion of withdrawn water that is not available for reuse, because it is lost to evaporation during the heating and cooling process in power plants, or because it becomes polluted and must be treated as waste water.

This great reliance on freshwater makes the energy supply chain, and power plants in particular, vulnerable to variations in the water supply. Indeed, we are already seeing cases of powerful energy companies losing the battle over scarce water resources. In a report published by the Union of Concerned Scientists, "Water-Smart Power: Strengthening the U.S. electricity System in a Warming World," Rogers et. al. note there were many instances between 2006 and 2012 of plants that had to reduce their output or shut down completely because available water supplies were insufficient (either in volume or temperature) to run the plant effectively.[5] In 2012, a plan to build a 1,320 megawatt coal plant in Texas was blocked by regulators, out of concern about the impact of the 8.3 billion gallons of water developers were proposing to siphon off the Colorado River each year to run the plant. Plans for the project were scrapped completely in 2013.[9]

Hydroelectric power plants are not thermoelectric. Instead of using steam to generate power, they rely on running liquid water to turn turbines directly. Water scarcity therefore has a rather obvious potential to limit the output of hydroelectric plants, and we have seen this play out during recent severe droughts in the western United States. A 1 percent reduction in the flow of the Colorado River, for example, can reduce energy output from the river's various hydroelectric plants by 3 percent.[10] In 2014, the California Energy Commission reported that drought would reduce the amount of power generated from California hydroelectric plants, and that the state would have to turn to natural gas to make up the shortfall, increasing emissions beyond what the state had planned.[11]

All of these pressures are likely to be exacerbated as the effects of climate change mount. As noted later in this chapter, climate scientists believe we can expect more severe and frequent droughts in the future due to climate change. Even under normal precipitation conditions, higher average temperatures will encourage evapotranspiration, and therefore lead to more irrigation for agriculture, creating a greater need for freshwater in the industry that is thermoelectric's greatest competitor for the resource.[8] Of course, it is not only water that is in greater demand in warm conditions, as space cooling needs also spike during these times. As temperatures rise, energy producers will need to find ways to produce more energy with less cool water.

Access to both fuel and water is being threatened by a lack of planning and absence of a policy that could help alleviate the strains on our supplies of fresh water. A Government Accountability Office report from 2012 found that: "Improved energy and water planning will require better coordination among federal agencies and other stakeholders. GAO's work has demonstrated that energy and water planning are generally 'stove-piped,' with decisions about one resource made without considering impacts to the other resource."[12]

We are in a time of transition in the energy sector, with inexpensive natural gas increasingly taking the place of coal power. Since the lives of power plants are generally measured in decades, the technological and policy choices we make for each new plant will have significant implications for the future of our energy system. Rogers et. al. write: "Understanding and addressing the water impact of our electricity choices is urgent business. Because most power sector decisions are long-lived, what we do in the near term commits us to risks or resiliencies for decades. We can untangle the production of electricity from the water supply, and we can build an electricity system that produces no carbon emissions. But we cannot wait, nor do either in isolation, without compromising both."[5]

ENERGY INFRASTRUCTURE

Periodically, the American Society of Civil Engineers (ASCE) releases a report card assessing the strength of American energy infrastructure, including evaluations of overall capacity, energy grids, other forms of distribution, and investment for the future. In 2013, the ASCE assigned an overall score of D+—the same grade our nation's energy infrastructure earned in the previous report in 2009.[13, 14]

Because our energy infrastructure was not developed as a cohesive national project, it has become an amalgam of generation and distribution facilities, strung together across regions as they have come online. Our energy system now includes approximately 5,800 major power plants, a network of over 450,000 miles of high-voltage transmission lines, and all of the local overhead lines or underground cables that bring power to each home and building.[15] Most of this infrastructure was developed decades ago, and some dates back to the 1880s. The ASCE has cited the age of this infrastructure as a primary factor threatening its reliability. The authors of the 2013 Report Card noted that instances of power outage rose from 76 in 2007 to 307 in 2011. Although most of these blackouts have resulted from weather events, a significant number of them were caused by "systems operations failures."

Aging power systems can lead to equipment failure, voltage surges, power quality issues, and blackouts or brownouts during periods when demand outstrips supply. ASCE argues that greater investment is needed to upgrade and coordinate energy infrastructure on a national scale. In the absence of any new programs, the organization predicts that investment in infrastructure will fall $107 billion short of where it needs to be by 2020, a deficit that will grow to $732 billion by 2040.[15] This means that as new power capacity comes online, and existing infrastructure ages, the investment needed to secure our national energy infrastructure will grow exponentially, eventually creating a hole so deep it's hard to imagine a reasonable means to climb out of it.

Interruptions to the power supply caused by aging infrastructure will take an economic toll. A failure to spend a reasonable amount now on infrastructural upgrades will result in far greater costs down the line. ASCE estimates that, by 2020, these interruptions will cost households and businesses $71 billion, a figure they predict will rise to $354 billion by 2040.[15] Ignoring for the moment the costs of cascading outages resulting from weather events, we can see great costs associated with energy equipment failure from the not-too-distant past. The August 2003 blackout in the Northeastern United States, for example, affected 50 million consumers and resulted in more than $6 billion in losses. Collectively, energy quality, interruptions, and outages cost the U.S. economy over $80 billion each year.[16]

Age alone does not account for the increasing fragility of our national energy system. This infrastructure is being called on to perform functions today that engineers could scarcely have imagined when much of it was created: widespread use of air conditioning, diverse plugloads ranging from phone chargers to MRI machines, and the high energy, uptime and power quality demands of operations like data centers. As our use of energy and electricity has evolved, infrastructure has lagged behind.

This is evident in the power grid itself, which features increasingly diminished generation and transmission capacity, and in the increase in bulk power transfers among regions.[17] The problem is also seen in how the grid is managed. For most of its history, the electric power grid was owned and operated by a network of private, regional and local utilities, each with their own geographic territory. Each utility operated as a regulated monopoly. There was little competitive pressure, and regulators ensured that the utilities were delivering power reliably, with minimal risk to the system. Deregulation (also known as utility restructuring) has allowed multiple utilities to operate in the same area, and has weakened the oversight and coordination that ensured reliability and adequate supply in the past. The result is what University of Minnesota Professor Massoud Amin has referred to as a less "shock absorbent" system.[18] The system has become more disorganized, and we have lost the planning and coordination that created a safe and reliable energy supply in the past. Amin writes, "As a result of these 'diminished shock absorbers,' the network is becoming increasingly stressed, and whether the carrying capacity or safety margin will exist to support anticipated demand is in question."[16]

Threats to our energy infrastructure are not limited to those affecting the generation and distribution of electricity. The process of extracting, transporting, and distributing fuel requires its own complex infrastructure, often stretching not just across the continent but around the globe. It is in this context that energy security is often discussed in national politics; that is, the need to decrease our dependence on foreign fuels as a means of securing our energy supply. This line of reasoning seems to assume that as long as fuel is extracted or harvested domestically, the energy security problem will be solved. We have indeed seen a boom in the development of North American energy supplies in recent years. Domestic natural gas production has grown from 17.93 trillion cubic feet in 1990 to 24.07 trillion cubic feet in 2013.[19] However, the rapid expansion of natural gas production capacity has brought its own incumbent infrastructure challenges.

Our nation's natural gas infrastructure now includes 325,000 miles of transmission pipelines and 2.15 million miles of distribution pipelines. Most of this pipeline is owned and maintained by private entities, which number in the hundreds. The U.S. Transportation Security Administration indicates on their website that "while pipelines are an efficient and fundamentally safe means of transportation, pipeline disruptions can have effects that impact the economy, and, in extreme cases, public health and national security." Indeed, despite recent improvements in pipeline safety, incidents involving natural gas pipelines still cause an average of 17 fatalities and $133 million in property damage annually.[20]

As vast as this pipeline array has already become, it is still insufficient to move the quantity of natural gas that is being generated domestically and to supply all potential markets within the United States. In a 2014 report, a team examining pipeline infrastructure at the Brookings institute noted that a large portion of the natural gas production that has come online in recent years is handled in areas that have not traditionally been associated with natural gas production (such as Pennsylvania), and where there is little preexisting pipeline infrastructure.[21] New pipelines are being developed quickly to transport this gas for sale, but this expansion does not affect every market. In the northeast, challenges associated with developing new natural gas infrastructure (including population density, environmental concerns, and lack of skilled labor) have been sufficient to prevent large scale deployment of new pipelines. In the end, it may be more cost effective to bring the natural gas to port for export than to attempt to reach domestic customers.[22]

Considering the documented leakage of natural gas from extraction facilities and pipelines, domestic natural gas starts to seem less like a panacea for American energy security. Methane is a potent greenhouse gas. Leaks from natural gas infrastructure represent the largest anthropogenic source of methane in the atmosphere. A 2013 study of a Denver-area fracking facility by the National Oceanic and Atmospheric Administration (NOAA) and the University of Colorado, found that 9 percent of natural gas extracted from the ground was lost to the atmosphere.[23] Average leak rates for fracking operations throughout the country are still in dispute,[24] but given the scale of the industry, any failure of infrastructure could lead to substantial leaks. As the report from Brookings notes, there is a lack of clear studies on the infrastructure risks for natural gas.[21] There is no doubt that natural gas represents a significant component of our diversified national energy portfolio, but infrastructural problems remain that prevent this energy source from guaranteeing our energy security.

Whether a failure occurs in the transmission of fuel or the distribution of electricity, it is likely that the effects will be felt broadly within our energy system. Our energy infrastructure is highly interconnected and interdependent, and was not designed to promote resiliency of the entire system when one part fails. This vulnerability is compounded by the interconnectedness of modern energy end-uses. Only a few generations ago, an energy bill was known as a "light bill" because, at least in households, electricity was used for little else. Today, nearly every function of society relies on a constant, reliable stream of energy. From banking to agriculture, transportation to government, nearly everything we do shuts down during a power outage, with effects of local outages rippling out across sectors and across the country. Any failure in our energy infrastructure can turn into a threat to our security, productivity, and economy. It is therefore particularly important to examine external factors that might affect these failures, climate change chief among them.

CLIMATE CHANGE

Climate change has already produced conditions in the United States that threaten our energy security, and these conditions are likely to worsen in years to come. As we have just discussed, our energy infrastructure is not particularly robust, and environmental changes can disrupt the delicate balance on which it relies. In a 2014 study, the Government Accountability Office found that "climate changes are projected to affect infrastructure throughout all major stages of the energy supply chain, thereby increasing the risk of disruptions."[25]

This general threat can be analyzed by looking at four main effects of climate change, and their implications on our energy system:

1. Increasing air and water temperatures
2. Decreasing water availability
3. Increasing frequency and intensity of storm events
4. Sea level rise

These effects of climate change are the same as those identified in a 2013 report from the U.S. Department of Energy titled "U.S. Energy Sector Vulnerabilities to Climate Change and Extreme Weather.[26] Of these risks, severe weather events have probably received the most attention, and they do pose a serious risk. However, other climate change-related factors (such as water scarcity, as we have discussed) will affect the energy system in a variety of ways—some more subtle than a hurricane, but many no less perilous.

Rising Temperatures

Perhaps the most obvious of these factors is increasing global temperatures. Average seasonal temperatures have been rising over the past century, and at an increasing rate.[27] As this book goes to press, 2014 was the hottest year ever recorded,[28] and all of the top ten hottest years since record keeping began in 1896 have occurred in the last 15 years. Rising overall temperatures have manifested as intense and more frequent heat waves, longer wildfire seasons, thawing and shrinking permafrost cover and sea ice, and a longer growing season.[29]

These conditions threaten the energy system in a variety of

ways. In the Arctic, temperatures are rising even faster than the global average, and the changing landscape is compromising oil and gas infrastructure. Oil and gas resources in Alaska are critical to the U.S. energy portfolio, but as permafrost thaws, it loses its ability to bear weight, imperiling pipelines, extraction equipment, and even buildings that have been erected to support the industry. The Alaska Department of Natural Resources limits the number of days per year that travel is allowed on the tundra to help protect its integrity, and over the past three decades, this number has fallen from 200 days to just 100 days, effectively cutting the amount of oil and gas activity that can occur throughout the year by half.[30] As we increase our reliance on northern territories for fossil fuel drilling, these environmental changes will make fuel more difficult, expensive, and dangerous to extract.

Rising water temperatures may limit the efficiency and effectiveness of power generation. Cool, fresh water is vital to every thermodynamic power plant, as it is needed to effect the phase change of steam back to water following the turbine. The pressure change between these two states drives the steam through the turbine, and any increase in water temperature decreases its ability to create this pressure differential. Increasing water temperatures also may make it difficult for certain power plants to meet regulations that limit the temperature of discharged water as means of protecting aquatic environments.[6]

Energy transmission may also be compromised by rising temperatures. The proportion of energy lost in transmission and distribution (currently about 7 percent) will increase as temperatures increase and the carrying capacity of the lines themselves decreases. Frequent and severe wildfires resulting from rising temperatures and drought are already destroying more remote energy distribution and transmission systems. These transmission systems will be strained further as rising temperatures increase the demand for energy used for cooling. We may see a decrease in fuel oil and natural gas used for heating during the winter, but peak demands in summer months are likely to increase, requiring higher loads.

Decreasing Water Availability

Warmer temperatures speed up the rate of water evaporation from the surface the earth and increase the capacity of the atmosphere to hold water. This seemingly innocuous change has widespread effects, disrupting regional water cycles and altering precipitation patterns.[31] Drought conditions have increased over the last 40 years, and 2014 saw a continuation of severe drought conditions throughout the Western United States, including what has become the worst drought in California's history.[32] Other effects of water cycle disruption include rises in rainfall in severe storms, more rain and less snow in cold climates, and earlier spring melts.[28]

As discussed earlier in this chapter, a decrease in the availability of fresh water may have serious consequences for our energy system. As climate change causes a rise in global temperatures, the energy sector is likely to find itself in greater competition for the water it needs to support its supply chain. Warm temperatures lead to increased demand for water for many of the processes we rely on for physical and economic health. Irrigation, livestock, and drinking reservoirs all require more water in warmer weather. Water supply will be threatened and become less reliable as temperatures rise.

Increasing Frequency and Intensity of Storm Events

The disruption in water cycles just described can lead to more frequent—and more severe—storms. As temperatures increase, the atmosphere can hold more water. Increased water in the atmosphere causes precipitation events to become more intense.[30] During the last 50 years, average rainfall in the most severe 1 percent of storms increased by almost 20 percent.[28] This increase in frequency and intensity will lead to greater damage from flooding, high winds, blizzards, tropical storms, and storm surge.[30] High precipitation in some areas means lower precipitation in others; droughts and heat waves have also become more frequent and intense in some regions.[29] These extreme weather patterns pose a wide range of threats to our energy system at every stage of the energy supply chain.

Much of the equipment used in the United States to identify new fossil fuel reserves, extract those fuels, refine and process them, and transport them for sale are located on or near a coast or offshore. This leaves equipment particularly vulnerable to tropical storm events. The Gulf Coast is central to the American energy system, and it is also particularly vulnerable to severe weather. Around 50 percent of our crude oil and natural gas production occurs in the Gulf Coast area, along with nearly 50 percent of its refining capacity. Energy infrastructure there is so dense, Hurricanes Katrina and Rita disabled or destroyed more than 100 platforms, damaged 558 pipelines, and forced many refineries to shut down for a period of several weeks.[29] Although Hurricane Sandy struck further north, it too did damage to our fuel production infrastructure, with nearly 7 percent of U.S. refining capacity in the direct path of the storm.[33] The Buckeye and Colonial oil pipelines were forced to close following the storm, accounting for 825,000 barrels of lost delivery per day. Sandy also halted shipments of fuel into New York Harbor, cutting delivery of fuel by nearly 60 percent in the days following the storm.[33]

Power production facilities are also vulnerable to storm damage due to their concentration along coastlines. Those plants closest to sea level face threats from floods during storm surges. Hurricane Sandy completely shut down several coastal power plants that were inundated by the storm. Further inland, power plants are often located adjacent to waterways for easy access to the water they need for production. As a result, they are also often located in low-lying flood plains, and are also susceptible to flooding.[29]

Energy transmission and distribution equipment is particularly vulnerable to severe weather. The number of incidents of grid disruptions from weather events appears to be on the rise (see Figure 1-3). Many of these power outages result from trees falling on one of the thousands of miles of distribution lines strung around the country. In more severe cases, the effects of a storm can cascade across our fragile and interdependent grid, leaving large regions in the dark. Hurricane Sandy affected 8.7 million power customers, 1.4 million of which were still without power six days after the storm.

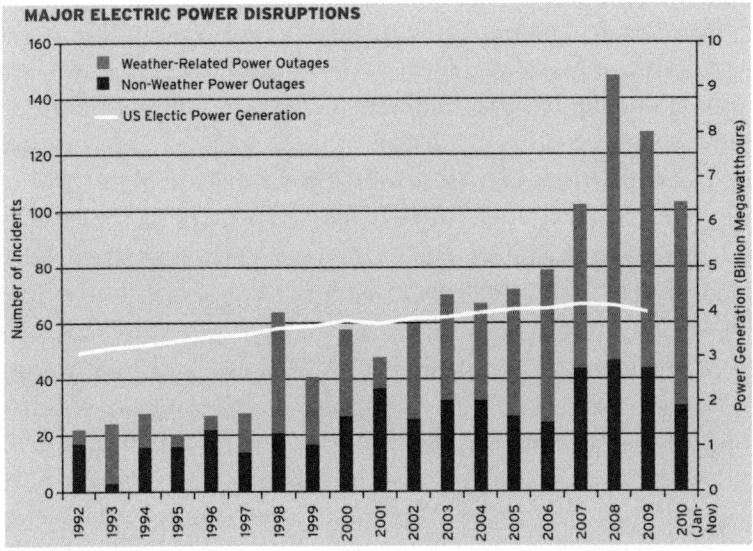

Figure 1-3: Weather Related Grid Disruptions, 2000-2012. Source: Chart by National Wildlife Federation, from data from North American Electric Reliability Corporation and the U.S. Energy Information Administration.[34]

The average global sea level has been rising at an increasing rate in recent years, following two millennia of little change at all.[35] As already noted, much of our country's key energy infrastructure is located near the coastline, in low-lying areas inland, or offshore. As the forces of climate change cause sea levels to rise, this infrastructure will be made increasingly vulnerable to flooding from heavy rains and storm surge. Severe storms in coastal areas can impact seawalls, levies and other infrastructure designed to hold back storm surge and rising sea levels, further exacerbating the risk to any energy infrastructure that becomes exposed.

SYSTEM VULNERABILITIES TO SABOTAGE

Reliable energy is the linchpin of nearly every economic and social function in our society. Finance, industry, agriculture, education, healthcare and governance all rely on the uninterrupted operation of our national energy system and its international supply chain. This dependency, and the damage that could result from a large-scale power disruption, may make the energy system a particularly attractive target for saboteurs. Potentially adding to this allure is the vast interdependency of the energy system itself, and the fact that a successful attack on a certain piece of the system is likely to cause cascading failures.

Seeking to better understand this vulnerability, the Department of Homeland Security asked the National Research Council (NRC) to study to matter. In 2005, the NRC convened the "Committee on Enhancing the Robustness and Resilience of Future Electrical Transmission and Distribution in the United States to Terrorist Attack," to assess the threats, develop recommendations for mitigating them, and issue a report to the government. This report was completed in 2007, but due to security issues, wasn't released to the public until 2012.[36] The NRC researchers describe an energy system in the United States marked by significant security exposures. These exposures fall into three general categories: those resulting from vulnerabilities in physical infrastructure, the digital systems that govern the infrastructure, and the personnel systems that run the energy system.

Physical Vulnerabilities

As described in the energy infrastructure section earlier in this chapter, our national grid was developed in parts and over time, without any kind of master plan for security. Energy generated at power plants flows through transformers, over long-distance transmission lines, into substations, and finally over distribution networks to end users. Historically, this electrical supply chain was controlled by vertically integrated utilities, but deregulation of the energy market, and the piecemeal undoing of that deregulation in certain states, has left the stages of power delivery system in different hands, and has made coordinated security planning very difficult. The interdependent nature of this system makes it vulnerable. If any one piece of the system is sabotaged, the entire system will fail.

In addition to this fragility, the power delivery system also has problems with resiliency. The severity of a power disruption is measured not just by how many people are affected, but by how long the disruption lasts. Certain kinds of equipment failures are much more difficult to fix, which can result in longer outages. Transmission lines, for example, are often hundreds of miles long and are therefore very difficult to secure. The destruction of just a few transmission towers could bring down many miles of transmission lines, requiring a major repair project before power could be restored.[36] High voltage transformers, once destroyed, could be difficult to replace. They are large and complex machines and are custom-built for each substation. What's more, the NRC report found that many kinds of transformers are no longer manufactured domestically, and that it can take months or even years to take delivery on a replacement.

Finally, it's important to reiterate that much of the infrastructure making up the energy system in the United States is decades old, with some pieces dating back 100 years or more. The general age of the infrastructure that undergirds the energy system in the United States does not just make it more vulnerable to failing on its own. It makes it more susceptible to sabotage. In addition to being largely unsecured, our energy infrastructure is not well protected against a disruption, and a small amount of tampering can leave the entire system out of balance, and therefore inoperable.

Cyber Vulnerabilities

Even as the physical energy infrastructure has aged, digital monitoring, communication, and automation systems have been built on top of it. Computer systems are now used throughout the grid to control equipment, analyze reliability, and manage the energy market. These are all potentially susceptible to sabotage,

Today's Energy Challenges

but the NRC report singled out supervisory control and data acquisition (SCADA) systems as the most vulnerable. These systems monitor substations and send control signals to other equipment, such as circuit breakers, to keep transmission and distribution systems in balance. Unlike damage to other computer systems, which is likely to stay contained, an attack on a SCADA system could have the potential to cause widespread outages, as entire transmission or distribution networks could be taken offline.[36]

The existence of a network that is not strictly local gives saboteurs an opportunity to affect these systems remotely, or even from another country. In June 2014, several cyber security firms discovered that a hacker group originating in Russia had been aggressively infiltrating the control systems of oil, gas and power companies in the United States and Europe. This group, known alternately as "The Energetic Bear," and "Dragonfly," used Trojan-horse hacks to gain access to the industrial control software used to manage the grid. It is assumed that the intent of these attacks was industrial espionage, but if the hackers were so inclined, the control wrested by the hackers could certainly have been used to cause major disruptions to power supplies.[37]

The NRC report concludes that although cyber attacks would not take nearly as long as a physical attack to repair, the threat is still a serious one. A cyber attack could also be combined with a physical attack to create a disruption larger than the sum of its parts. For instance, a cyber attack could be used to obfuscate a physical attack at a remote location, thus delaying the response.

Personnel Vulnerabilities

If physical infrastructure and computer systems form the first two layers of the energy system, human workers are the third. This final layer comes with its own distinct vulnerabilities. The energy system relies on a complex assemblage of organizations and individuals to manage generation, transmission, and distribution systems, maintain lines, service equipment, and manage the market, among many other roles and responsibilities. These roles no longer typically fit within one vertically integrated utility. To safeguard our power delivery system, we depend on people, with various competencies and levels of training, working at hundreds of organizations located all over the country and the world.

As the first line of defense, it is essential that all power industry personnel are properly vetted to ensure they have no intent to harm the power system and are properly trained to respond to emergencies. Although the personnel who manage and operate the U.S. energy system have an exemplary record, the NRC report found that more could be done to ensure the quality of background checks and training programs. For example, background checks are often outsourced to a service security company and are managed separately by each organization at work on the energy system. The NRC report recommends standardizing the credentialing process to ensure uniformity and quality, and suggests further that government clearance be required for workers who will be called on to share information with government and law enforcement personnel in an emergency.

The secure operation of the energy system depends on the highly trained, skilled personnel that operate it. This workforce, however, is aging, and a great deal of expertise will retire with them as their careers conclude. Add to this the fact that competition has driven down the size of the power industry workforce in the past few decades to address competitive pressure. What we are left with is a smaller, less-experienced workforce and a significant loss of institutional knowledge.[36] It is therefore vital that the energy sector find ways to recruit well-educated, well-vetted candidates as the next generation of power system professionals and give them high-quality training on the secure operation of the system, including how to respond to potential threats or emergencies.

The intent of this chapter was to provide some concrete detail regarding the threats to the energy system we rely on. For the purposes of this book, it is important to remember that if any of these threats are realized, the end result will be the same: the lights go out. The circumstances described in this chapter pose a real and credible threat to constant, reliable power supplies.

Recognizing the multiplicity of factors that can affect our energy stability, it is incumbent upon managers at each local institution to take action—to develop a plan for keeping their institution resilient in the face of these threats. There are ways to pursue energy resilience in ways that are not only cost effective, but also support other organizational goals. Chapters 3 and 4 will describe this planning process, and Chapters 5 through 7 will focus on the three main pillars of energy resilience and how to benchmark and improve your performance in each.

Chapter 2

Local Energy Resilience

ENERGY RESILIENCE VS. ENERGY INDEPENDENCE

For decades, the national conversation around U.S. energy security has been focused on the concept of national energy independence from foreign sources of oil. The ideal outcome, it was thought, would be achieved when no fossil fuels had to be imported into the United States for energy production, as price spikes and foreign control over supply were seen as clear threats to our national and economic security. Recently, however, Americans have begun thinking about energy security in new ways.

In 2008, the tech pioneer and former CEO of Intel, Andy Grove, made waves with an op-ed in the *Washington Post* by challenging the energy independence orthodoxy and introducing the term "energy resilience" to the general public.[38] In an expanded version of this essay published later that year, Grove wrote, "As national policy, we must protect the U.S. economy from interruptions in the supply of such a critical commodity [oil]—whether those interruptions are related to natural or political causes. I believe that the appropriate aim is to strengthen our ability to adjust to such changes—to strengthen our energy resilience."[39] Instead of seeking to achieve complete energy independence—a lofty goal that a succession of seven presidents had sought and failed to achieve—Grove argued that it is more practical to orient our energy strategy around an ability to withstand and adapt to a variety of shocks to the energy system. Instead of failing to achieve energy independence, we could instead succeed in building true *energy resilience*. The term "energy resilience," as employed in this book, is a spectrum between the endpoints of self-sufficiency and complete dependence on outside sources of energy.

THE IMPORTANCE OF LOCAL ENERGY RESILIENCE

In our analysis, we take Grove's general concept of energy resilience, and apply it to the local rather than national level. Although energy resilience is an idea that belongs in national policy discussions, it is most fruitful, and most critical, to pursue resilience on a local level. When the next Hurricane Sandy or Katrina hits, we should not expect that our regional energy grids will be any more reliable than they are now. It will be up to each locality, and each local institution, to make a plan for what they will do when the power goes out. In July 2013, the Department of Energy (DOE) published a report titled "US Energy Sector Vulnerabilities to Climate Change and Extreme Weather," which explains that not only is the national energy infrastructure not getting any stronger, the threats against it are growing worse:

"Increasing temperatures, decreasing water availability, more intense storm events, and sea level rise will each independently, and in some cases in combination, affect the ability of the United States to produce and transmit electricity from fossil, nuclear, and existing and emerging renewable energy sources... Federal, state, and local governments and the private sector are already responding to the threat of climate change. These efforts include the deployment of energy technologies that are more climate-resilient, assessment of vulnerabilities in the energy sector, adaptation planning efforts, and policies that can facilitate these efforts. However, the pace, scale, and scope of combined public and private efforts to improve the climate preparedness and resilience of the energy sector will need to increase, given the challenges identified. Greater resilience will require improved technologies, polices, information, and stakeholder engagement."[30]

Commenting on the report to *The New York Times*, the agency's Deputy Assistant Secretary of Energy for Climate Change

Policy and Technology, Jonathan Pershing, reinforced this point: "We don't have a robust energy system, and the costs are significant. The cost today is measured in the billions. Over the coming decades, it will be in the trillions. You can't just put your head in the sand anymore."[40] As these statements make clear, even if we were to achieve the ideal of energy independence on a national level, our own aging energy infrastructure would leave us without the security that independence is supposed to ensure. Short of waiting for Congress to initiate an Apollo-scale project to renovate and restore national infrastructure, the best bet for local administrators is to pursue energy resilience on-site. The vast diversity of environments and infrastructure in the United States also mean that each locality will face different threats and need to focus on different solutions to achieve energy resilience.

In a speech on climate change to Georgetown University on June 25, 2013, President Obama addressed the issue, noting, "What we've learned from Hurricane Sandy and other disasters is that we've got to build smarter, more resilient infrastructure that can protect our homes and businesses, and withstand more powerful storms. That means stronger sea walls, natural barriers, hardened power grids, hardened water systems, hardened fuel supplies."[40] It could be argued that the President omitted a key point: that the best way to harden power grids is to keep them local. In the same speech, he praised New York City for engaging in a project to "[fortify] 520 miles of coastline as an insurance policy against more frequent and costly storms."[40]

As detailed later in this book, city-wide infrastructure upgrades are an attractive and practical response to a lack of broader action toward resilience. However, city-wide action may be similarly slow to come, and institutions within a city as big as New York may not want to rely on actions taken by the municipality to safe-guard their vital services. As several of the case studies in this book will illustrate, there were institutions in New York before Sandy hit that had already incorporated this perspective into their planning, and benefitted from their pursuit of resilience at a local level. Even within an institution, we can see how different approaches toward resilience led to different outcomes during Sandy. As much as NYU, for example, was lauded for keeping the lights on in many of its residential and administrative buildings throughout the storm and its aftermath,[41] the NYU Langone Medical Center was heavily criticized when it was forced to evacuate its patients during the storm after losing power.[42] Energy-intensive research buildings at the University were described as suffering "inestimable losses" in the storm, including live animal research subjects.[41] Those areas of the university that maintained power benefitted from an on-site natural gas-powered co-generation plant and microgrid (more on this technology in Chapter 6) that took over when the regional grid was knocked out. In contrast, the hospital and many research areas relied on back-up generators that failed or ran out of fuel.[41, 42] Different units of the university were at different places along the energy resilience spectrum when the storm hit, and consequently experienced dramatically different outcomes.

FINDING A PLACE ON THE ENERGY RESILIENCE SPECTRUM

As is starkly illustrated in every severe storm and regional blackout, each community's place along the energy resilience spectrum is determined by the response to the question, "What happens when there's a shock to the regional power supply?" In other words, how resilient are the operations of each community to the energy interruptions they will inevitably face? When a storm knocks out the regional grid or energy prices suddenly skyrocket, how many of the vital services that are performed within each community are able to continue and for how long? What percentage of the population that lives within will have their needs met and be adequately cooled, warmed, and fed?

Ambitious administrators and facility managers sometimes set their sights on total energy independence when confronted with the threats posed by an insecure regional grid.

However, with the exception of a few specific cases—remote, forward-positioned military bases, for example—complete energy independence is rarely the best management strategy in terms of capital costs, maintenance costs, or even energy security itself. Instead, the best target for each community can likely be found at some point along the spectrum of energy resilience and must be determined on a case-by-case basis.

Figure 2-1 displays the relative costs associated with a variety of scenarios. This energy resilience spectrum illustration is not based on a quantitative analysis of system costs, but represents the relationships between investment and risk identified in the authors' research and experience.

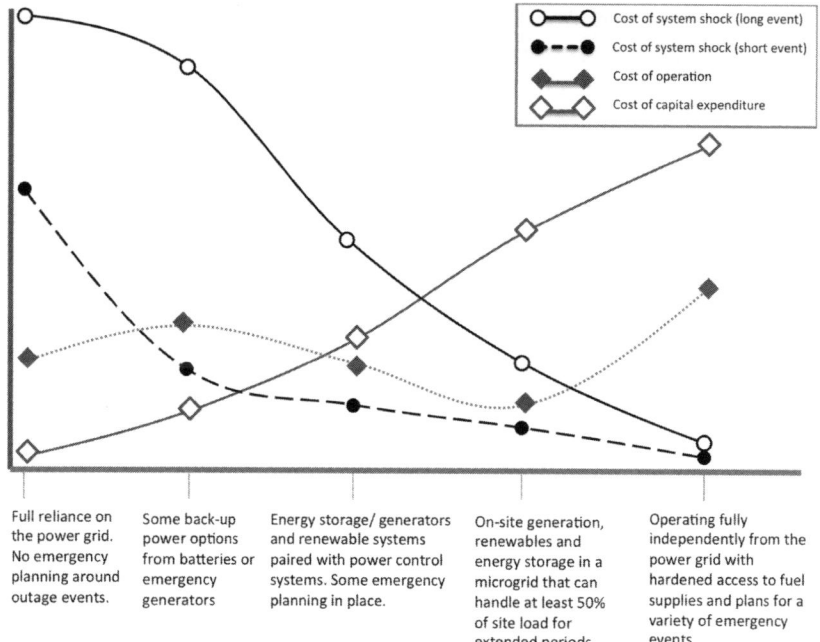

Figure 2-1: The Spectrum of Energy Resilience.

As Figure 2-1 illustrates, the cost of system shocks (be they outage events, supply challenges, or price spikes) can far outweigh the regular cost of energy. Taking even the most basic of energy resilience measures can dramatically reduce the financial impact of shorter system events.

At a certain point along the spectrum, institutions may find that pushing ever further toward energy independence may involve complex facility upgrades or generation equipment that is expensive both to install and operate. In these cases, institutional administrators must weigh additional expense against the cost of potential threats to system reliability. For some communities, going to the far right of the resilience spectrum will make sense. For others, the optimal outcome may be in the middle or even closer to the left.

Just like energy resilience, environmental sustainability can be imagined as a spectrum. Sustainable design and operation are not an all-or-nothing game, and every decrease in pollution and increase in efficiency can improve environmental outcomes. However, it is important to note that although energy resilience is often associated with greater environmental sustainability, either can be achieved without necessarily affecting the other. In November 2012, *The New York Times* quoted the urban planner and developer Jonathon Rose describing the dichotomy this way: "After 9/11, Lower Manhattan contained the largest collection of LEED-certified, green buildings in the world, but that was answering only part of problem. The buildings were designed to generate lower environmental impacts, but not to respond to the impacts of the environment."[43] Unless the design of a facility included a plan for protecting the flow of energy during a system shock like Sandy, an otherwise sustainable building could find itself in a vulnerable position on the energy resilience spectrum.

Without a well-known benchmarking system like LEED and third-party guidance from organizations like the US Green Building Council that developed the LEED standards, it can be difficult for administrators and facilities managers to plot a course toward energy resilience and to find a place on the energy resilience spectrum that is right for their community or institution. The remaining chapters of this book are designed to guide you through this process, and to familiarize you with the technologies and techniques involved. In each of the relevant management areas (energy efficiency, on-site generation, and emergency planning), we provide ways for managers to assess their current performance and plot a practical course to greater energy resilience. However, before this planning process can begin, it is important to understand that the right position for each community will be determined by its unique operational and financial circumstances and concerns. A higher position on the energy resilience spectrum is not always better.

Given our definition of energy resilience as the relative ability to protect the functioning of vital operations during interruptions to the regional power supply, it is important that each institution prioritize its functions and operations in terms of how much energy is required and how critical each function is to human safety and to the mission of the institution. This prioritization can then be weighed against the capital costs and payback of the energy resilience projects that would address those needs. For instance, a university typically includes student housing (where electricity, lighting, and heat are vital), administrative offices, classroom buildings, research facilities, and athletic centers. If the university invests in on-site generation and a microgrid, but cannot generate 100 percent of the campus's typical load, it will have to make decisions about which facilities receive power in emergency situations. This may include lighting and heating the dorms and temperature control at research facilities. It may only include the athletic center, which can then be used as a refuge.

This is an oversimplified example of a complex process, but it illustrates that more is not necessarily better when it comes to energy resilience. Unlike energy independence, energy resilience is all about finding that sweet spot where an institution's ability to protect its vital functions from power outages converges with sound financial planning.

The financial benefits of an energy resilient system go beyond day-to-day operational savings—they also include the avoided impact of major downtime during outage events. Delivering a resilient, uninterrupted energy supply to a community's most critical areas and functions can prevent property damage, avoid liability, protect the health and safety of citizens or residents, and preserve mission through unimpeded operations. The case studies in this book illustrate how administrators and managers throughout the country have adopted the process of finding a place on the energy resilience spectrum, and describe the benefits they have realized.

Of course, projects that improve energy resiliency often improve financial security as well. Two of the key components of most energy resilience plans—on-site generation and operational efficiency—are often associated with significant opportunities for energy cost savings. As detailed in the following case study, the Gunderson Health System in Wisconsin was able to shave 25 percent off its annual energy bill by retrofitting its facilities for energy efficiency and switching to on-site generation wherever possible. This is no small feat when you consider that Gunderson had been spending $5.3 million a year on energy.[44] By purchasing cheaper energy from on-site generation and less energy overall thanks to energy conservation measures, communities can significantly improve financial security while ensuring energy resilience and helping to protect their overall mission. Later chapters will outline technologies and strategies in the areas of energy efficiency and generation.

In this situation, administrators found a place on the energy resilience spectrum that worked for the institution. Often, we hear of the ability to go into "island mode" (a term used to describe the ability to power operations with on-site generation when the regional grid is down) as the gold-standard of energy resilience. As this case study shows, there are many ways to measure and achieve resilience that don't necessarily include "island mode."

Case Study—Gundersen Health System's 'Envision' Program[44]

Sprawling across three states, the Gundersen Health System encompasses five hospitals, 24 regional clinics, and dozens of additional local clinics specializing in everything from eye care to orthodontics. It represents the kind of grand bureaucracy that is sometimes associated with an inability to take the kind of bold, future focused actions necessary to pursue energy resilience, especially when those actions are associated with large capital expenditures. Gundersen, however, is defying that stereotype, and in the process, helping other hospitals and healthcare organizations see that planning for energy resilience is not just responsible citizenship, it's good business.

In 2008, the health system was confronted with a harsh reality in the form of a $5.3 million energy bill. Hospital administrators knew that if they did not act quickly to reduce this cost, it would inevitably lead to higher prices for their healthcare services. Gundersen set about retrofitting their buildings and managing staff practices in an attempt to control healthcare costs, and they saw benefit from these efforts fairly quickly. It was at this time, as Jeff Rich, the Executive Director of Gundersen's environmental and energy program, explains, that the administration of Gundersen decided to take their efforts to the next level. "We had a lot of momentum right out of the gate for our energy efficiency efforts. Once we got going, the question we kept asking ourselves is, what can we do next? This led to the creation of the Envision program."

Gundersen created a dedicated team devoted to the pursuit of energy resilience and environmental sustainability. The program they created was given the name "Envision." Although the Envision brand now includes all of Gundersen's environmental efforts, including things like waste management, sustainable foods, and transportation the core of the program is still in energy management. By 2013, Gundersen had cut its system-wide energy bill by $2 million, or almost 40% under 2008 levels. Although a new hospital at Gundersen's La Crosse campus, which opened in January 2014, initially added to the energy load, the plan is to completely offset new demand with energy efficient design, geothermal heating and cooling, and on-site power generation.

Rich explains that regulatory and technical hurdles have prevented Gundersen from taking advantage of energy efficiency and on-site generation at the new hospital to go into "island mode" when the regional grid blacks out. However, he believes that these measures, and similar measures taken throughout the Gundersen system, have increased the energy resilience of the facilities. For

> instance, all heating and cooling for the new hospital at La Crosse will be handled on-site, with a network of geothermal wells under the parking lot, and with chiller/tower optimization, thereby greatly reducing the demand on back up generation in the event of an interruption to the regional power supply. "More importantly," Rich explains, "all of the energy efficiency advancements we've made through the Envision program have made us more resilient to future volatility in the energy market. Natural gas is cheap now, but may not be forever. Investments that we make in energy efficiency ensure that we will be able to control and predict our energy costs, and therefore control and minimize the cost of our healthcare services."
>
> The Envision program has been so successful for Gundersen, that they have begun to offer consulting and training services to other hospitals and healthcare institutions interested in following in their footsteps. "We offer energy consulting and training services under the Envision brand as a way to ensure that the strategies we have developed have a maximum impact. We don't see ourselves as an engineering company, but we want to educate others and help them advance. We started pursuing energy efficiency as a way of keeping costs down for patients, but we have an environmental and health mission that extends beyond our walls."

ENERGY RESILIENCE IN OUR FOUR COMMUNITY TYPES

To illustrate the advantages that local communities and institutions have in pursuing energy resilience, we have organized the discussion in this book around four key community types: small municipalities, educational campuses, healthcare campuses, and military bases. Each of these community types has distinctive characteristics that give them an advantage in achieving energy resilience, compared to a single building or a larger community. These community types share four key characteristics that make energy resilience activities more cost effective:

1. **Place**: Each of these communities is defined by a contiguous footprint that allows for the coordination of energy resources for multiple loads without long-range transmission lines.

2. **Permanence**: These communities cannot simply relocate. For various reasons, where they are now is where they will likely be in 50 years. This makes long-term facility and infrastructure planning feasible.

3. **Control**: The governing bodies of these communities have control over (at least a portion) of the infrastructure and processes that drive energy usage in the community. This control allows for long-term energy planning.

4. **Mission**: All of these communities have facilities that play a role that is central to the mission of the organization. Whether it is a fire station, a student dormitory, an operating theater, or a satellite control room—continuous energy supply is essential to carrying out the vital mission of each community.

Place: Control Over Land and Local Infrastructure

First, the land on which a community sits can offer a built-in advantage for the development of renewable energy capacity. Access to land has been found to be one of the most important barriers to the expansion of renewable energy production.[46] The development of nearly any kind of local power generation will require available space and must meet certain usage regulations. Combined heat and power plants must conform to emissions rules; wind turbines need to meet height, noise, and wildlife protection restrictions, and even photovoltaics and solar water heaters face space-use challenges based on safety and aesthetics. Among the few locations that may meet all of these restrictions, some may still face a high degree of competition for land with agricultural, recreational, or development interests.[46] Institutions that can develop energy infrastructure on wholly owned land can avoid this competitive pressure, and some regulatory hurdles. The concept of *place*, as we use it, also describes the local setting itself. The fact that the institutions we examine occupy a single, relatively small geographical space provides them with built in advantages in terms of energy resilience, especially related to on-site energy generation, as we will describe in depth in Chapter 6. When energy is generated at the same site as the institution that will consume it, the infrastructural vulnerabilities that were described in Chapter 1 are significantly reduced. Since this generation is under the control of the community itself, it can be carefully matched to specific loads it is intended to meet. Personnel can carefully maintain the local energy system to ensure reliability, and create emergency plans that cover the entire energy life-cycle, from generation to end-use. The local setting allows for the kind of control that is necessary to achieve true energy resilience, as described below.

Permanence: Opportunity for Long-term Planning

The space that each of these communities occupies tends to be its permanent home. A domestic military base or hospital may eventually close, but building these institutions takes such a large investment, they are usually intended to stay open indefinitely or at least for several decades. Obviously, the closing and abandonment of a university or municipality is even less common. At colleges and universities, building assets are typically owned by the institution and managed for the same general purpose for many decades.[47] Most communities, including small towns, plan to operate at least 50 years into the future.[47] This permanence allows for long-term planning and helps justify infrastructural investments, such as energy generation equipment or energy efficiency upgrades. At all of our four community types, it often makes operational and financial sense to invest in upgrades that will return an energy resilience benefit in the short term, even if that financial benefit is not realized until the mid- or long-term planning horizon.

Control: Centralized Oversight

Central governance, or control over the energy processes within a community, is another key characteristic that links our four community types. Decision makers must have sufficient authority over energy purchase and generation, building infrastructure, and end-use practices to engage in effective energy resilience planning. As noted in Chapter 2, this doesn't mean that all decision-making power has to reside with one individual or within a small group. It is likely that successful energy resilience projects will result from the coordination of multiple departments (maintenance, finance, administration, etc.) and groups of stakeholders. What's important is that it is ultimately within the authority of the institution to perform this coordination, and to take action on ambitious energy resilience projects. This authority is what allows for investments in on-site generation equipment, energy efficient technologies, continuous commissioning, and microgrids. This level of control can also make it easier for administrators to manage the demand associated with existing infrastructure. In a large, municipal setting, planners must apply regulations, zoning restrictions or incentives to spur individual building owners to retrofit their homes or businesses for energy efficiency. In a centrally governed institution, these efforts can be mandated and coordinated. When buildings and infrastructure are owned by one organization, a community has the data and control it needs to create and implement a coordinated energy plan.

Mission: Safety and other Essential Functions

Our four community types share elements of their core mission that make energy resilience a priority. First, each community type holds the safety of their constituents as a primary concern. Here, we use the term "constituents" to mean residents at a military base or residential community, students at an educational campus, and patients at a hospital campus. Energy resilience efforts help ensure safety by protecting the systems that support it, such as heating/cooling, catering, life-support equipment, and security systems. Second, it is fair to assume that cost-effective operation is a common priority among these (and perhaps all other) community types. As discussed in depth further in this book, energy resilience measures can go far to protect institutions from financial risk. Finally, we define energy resilience as the ability of an institution to shield its core mission from shocks to the energy system. All four community types are associated with an especially vital core mission. Whether educating students and conducting sensitive research, caring for the ill and injured, housing and protecting residents, or defending our national security, the mission of each of our four community types must be carried out every day of the year, and cannot tolerate interruptions from an extended power outage.

The qualities listed above make each of our four community types particularly well suited for energy resilience projects. In the remainder of this chapter, we will briefly discuss the particular need of energy resilience at each community type, and the kinds of things that are being done at each to meet this need.

Military Bases—The Energy Resilience Imperative

The importance of energy resilience on military bases all comes down to the fourth characteristic listed above: *mission*. Like any institution, energy resilience on a military base means the ability to continue vital functions uninterrupted by shocks to the regional power supply. Unlike most other institutions, the mission that these functions support is to protect our national security and the safety of those in the armed services.

The Department of Defense is the largest consumer of energy in the U.S. Government, and a great deal of this energy is consumed at military bases. These installations consume about 3.8 billion kilowatt-hours of electricity each year, or enough power for 350,000 U.S. households over the same period.[48] The operations supported by and housed within our military installations are vital to our national security in countless ways, both seen and unseen. From the protection of proprietary technologies and storage of sensitive data, to training and preparing our fighting forces, to housing and feeding their families—every activity on a military base is critical in its own way. The importance of keeping security operations unhindered is elevated in emergency situations, when a disruption in the grid may be more likely.

If the demand for energy at these facilities goes unmet, even for a short time, the operations and the imperatives they address could be compromised. In addition to the enormous demand for electricity, energy security on domestic military bases is further imperiled by the fact that most are located in remote locations, at the end of long transmission feeders.[49]

Although we focus primarily on domestic bases in this book, it is important to note that the exposure of domestic bases to interruptions in energy supply is relatively tame compared to the risk that affects our bases abroad. The energy intensity of our forward-positioned military installations is a serious constraint to our operations.[50] The strategic options open to our military are limited by the complex logistical demands associated with delivering fuel in the field. Military bases can only be placed in the field where it is logistically feasible to deliver the fuel needed to operate that base and run the operations that it will support.

Military Bases—The Energy Resilience Opportunity

The characteristics of *place, permanence,* and *control* are at the center of the US military's response to the energy resilience imperative on base. Military management has the *control* necessary to take action on this issue, and there is little doubt that they understand the challenges posed by reliance on the commercial grid at home and far-flung distribution networks abroad. It was this understanding that led the Department of Defense (DoD) in 2008 to establish a task force to tackle the problem of energy use in U.S. military installations, dubbed "Net Zero Energy Installation" (NZEI).[51] This group included delegates from the office of the Secretary of Defense, all four military services, the Federal Energy Management Program (FEMP) at DOE, and the National Renewable Energy Laboratory (NREL).[48]

It was an early priority in the task force to explore the possibility of achieving net-zero energy status on a military base. To earn this label, a base would have to generate all the power it required with on-site, renewable sources.[48] Marine Corps Air Station Miramar was chosen as a test case for this concept, given its already strong record for aggressive energy management. In partnership with the National Renewable Energy Laboratory (NREL), DoD carried out a top-to-bottom study of the base, its operations, and opportunities for energy efficiency and local generation. This analysis became a template for the pursuit of energy independence at other bases—some of which are profiled elsewhere in this book.[48] This kind of coordination is an excellent example of the kind of *control* and coordination necessary to effect energy resilience.

In choosing military bases as the setting for their pursuit of energy resilience, the U.S. military can take advantage of the unique characteristics of *place* typical at these sites. NREL is now working on energy projects at 60 different DoD installations, and is pursuing complete net-zero energy status at two bases: Fort Carson, in Colorado, and Fort Bliss in Texas.[52] Although physical isolation is a liability when relying on the commercial grid, it is proving to be a boon in efforts to remove this reliance.

Since DoD often owns a significant area of land surrounding a military base, no new acquisitions are necessary to house local generation infrastructure, such as solar photovoltaics, and no long-range transmission lines are required. This is helping solar PV to approach grid parity prices for applications at military bases.[49] At Fort Carson, solar arrays are already generating three megawatts of electricity, or enough to power 800 homes on the base. Fort Bliss has installed hundreds of PV panels, and officials there are hoping to start work soon on additional arrays that would increase output to 20 megawatts. This would still not be enough to meet the enormous energy needs of the base, and administrators are also looking at geothermal heating and cooling (to offset energy usage) and wind turbines to supplement their renewable generation on site.[49]

Permanence is another common (though not universal) quality of military bases that makes them well suited for energy resilience projects. The military bases we focus on in this book, including those that are the subject of case studies, are permanent, domestic bases, intended to operate into the foreseeable future.

These bases are centrally managed by an organization capable of planning on a long-term time horizon, so investments can be made now toward the pursuit of future goals. As Fort Carson's Utilities Manager, Vince Guthrie, put it to *The Denver Post*, "What this program is about is security—economic security, environmental security and national security."[53] The department's FY 2013 budget included $1.1 billion for investments in conservation and energy efficiency, and almost all of that is directed to existing buildings.[51] That's a significant sum, even when compared to the $19.3 billion DoD spent on energy in FY 2011, $4.1 billion of which was spent on powering facilities.[51] It takes a particularly long view to spend a full 25 percent of an annual energy bill on energy reduction programs, especially when this represents billions of dollars. Energy resilience at military bases is pressing enough to justify this cost, and thanks to the administrative control at DoD, the decision can be made to address this need.

Healthcare Campuses—The Energy Resilience Imperative

As with military bases, the *mission* of hospitals and healthcare campuses is essential to understanding their need for energy resilience. Energy security can literally be a life-or-death imperative for hospitals, and energy costs often make up a significant portion of their overall budget. Healthcare campuses are among the most energy intensive facilities in the country, and, combined, spend more than $8 billion on energy every year.[50] As the DOE points out on its website, that amount of money could cover the salaries of more than 100,000 registered nurses.[50]

When significant power demand goes unmet, damage can accumulate quickly. The case of Langone Medical Center, mentioned earlier, is a stark illustration. As Hurricane Sandy settled over New York last year, the East River swelled and poured into the basement of the hospital. Although most of Langone's back-up generators were located on high floors, the fuel tanks that fed them were in the basement. When they became submerged, they either detached from the fuel lines or were automatically shut off when their liquid sensors detected the flood.[54] The power was out and, during the next 13 hours, the hospital was forced to evacuate 300 patients. It also suffered more than $700 million in damage, much of which is attributable directly to the outage (lost revenue, interrupted research, and paying employees who were not able to work).[55] Though it may have attracted unwanted attention due to its situation in the middle of a high-profile storm and city, Langone is hardly alone in this hardship. In 2012, 23 percent of the hospitals inspected by the Joint Commission, a healthcare facility accreditation group, were found to be out of compliance with standards for backup power and lighting.[54]

Hospitals and other healthcare campuses serve a vital role in our society, and each must pursue a proactive energy resilience program to ensure that they are able to reliably carry out their missions into the future. This imperative is increasingly well recognized in the industry, and in 2009 the DOE established the Hospital Energy Alliance to guide healthcare facilities through the process of becoming more energy efficient and resilient.

Healthcare Campuses—The Energy Resilience Opportunity

As a founding member of the Hospital Energy Alliance, Gundersen National Health System (see full case study earlier in this chapter) is lending its expertise in the area to other hospitals wishing to follow in its footsteps. Gundersen's success (25 percent reduction in energy costs amounting to $1.3 million per year, and greater resilience from lower demand and redundant energy systems) has been realized across a large network of hospitals and clinics that span the Upper Midwest.[50] However, instead of providing a counter to our conclusion that energy resilience is best tackled at a local level, the Gundersen case actually reinforces this point, since all of the achievements have been predicated on a local, facility-by-facility approach. The success of the Gundersen case is based on the characteristics of *place, permanence,* and *control* we have described.

Like a military base, healthcare campuses are typically permanently located on owned, contiguous space, allowing managers to plan and implement energy projects that require these characteristics of *place*. Many of these projects make use of the existing land and infrastructure of the hospitals themselves, such as the hundreds of photovoltaic panels that now cover the parking garage at Gundersen Lutheran Hospital. The need for extra land and infrastructure has also led Gundersen to seek partnerships with other community organizations to further expand its on-site generation capabilities. Gundersen Lutheren features wind power built on land and owned by a partner, and an on-site combined heat and power plant fed by biogass from a nearby landfill.[44]

To ensure on-site heat and power generation is able to meet demand in the most efficient means possible, hospitals must institute energy efficiency measures throughout the institution. The implementation of these measures relies on the strong central *control* typically seen on a healthcare campus. At the University Medical Center of Princeton at Plainsboro, in New Jersey, which opened in 2013 (see full case study later in this book), careful management and planning allowed the entire hospital and all of its operations to be powered with on-site generation. Instead of using its on-site systems to back up power from the regional grid, Dell actually uses grid power as the backup. Should both systems fail simultaneously, there are battery arrays that provide a third backstop for essential equipment.[45] This accomplishment is the result of a systems approach that is only possible if all components are planned in concert, and carried out with central oversight and control.

Educational Campuses—The Energy Resilience Imperative

Every college or university campus is like a small city unto itself, with a wide array of functions and facilities, and all of the attendant energy requirements. The importance of energy resilience on college campuses again comes down to *mission*. Residential students need HVAC, hot water, light and electricity in their residences, and cafeterias capable of serving three hot meals a day. Colleges cannot simply close if the power goes out and, as we discussed earlier, the provision of a safe and secure environment for student residents is central to the mission of any residential campus. The mission of many universities also includes ongoing research, much of which must be powered without interruption to preserve experiment parameters.[56]

On the same night that a shortage of oil for backup generators caused an evacuation of NYU Langone Medical Center, several of the university's research centers lost power, including the Smillow Research Center, which houses long term live animal research projects on heart disease, cancer, and neuroscience. Freezers thawed and animal areas flooded, causing years of research to be lost. "It is not very different from someone losing their entire home. For scientists, their research is their lifeline," Dr. Dafni Bar-Sagi, NYU's Senior Vice President for Science and Chief Scientific Officer told a reporter. "For someone who started three to four years ago, and just got to a point to launch their research program, it's time to rewind and start from fresh."[56] Meanwhile, when power was lost at many of NYU's lower Manhattan dorms, they too had to be evacuated, and students moved to alternate, often non-residential university buildings where they could be kept warm and fed.[57] Although Hurricane Sandy certainly highlighted the importance of energy resilience on campus, many colleges and universities had already sought to improve their energy performance as a means to protect their mission.

Educational Campuses—The Energy Resilience Opportunity

Driven by the need to control costs and a desire to promote environmental sustainability, energy efficiency is being pursued to some extent at nearly every college and university in the country. Many schools are realizing that to truly protect their mission, they must go beyond efficiency to achieve broader energy resilience. These campuses take advantage of the characteristics of *place, permanence* and *control* to find success in energy resilience. *Permanence and place* are key to Cornell University's strategy to become energy resilient, and the university is now able to meet its entire peak load of 35MW with on-site generation (mostly from a CHP plant which also produces 90 percent of the university's heat) and its cooling needs with a deep water lake-source cooling facility. When the power from the regional grid goes out, the university will continue to operate uninterrupted.[58] The school is also planning on taking advantage of its rural setting to install 6,766 solar photovoltaic panels on a 10-acre site they own.[59] Even in urban environments, colleges are finding ways to take advantage of their centralized *control* over infrastructure to plan large-scale energy projects that have the potential to make them more energy resilient. Lacking the surface space enjoyed by Cornell, NYU installed a CHP plant underground. While strolling through the plaza at 251 Mercer Street, a visitor would never know that beneath their feet lay a state-of-the-art, $125 million CHP plant that took over two years to build. It provides electricity to 22 NYU buildings, heat and hot and chilled water to a total of 37 buildings, and without it, NYU would not have come through Hurricane Sandy as a success story.[60] Although it may not seem like much compared to the capital cost, the $5-8 million in energy cost savings NYU is realizing from the plant is considerable.[60] Still, it took a central administration with a tolerance for long-term payback to make the project a reality and to realize the enormous energy resilience benefits it affords.

Residential Communities—
The Energy Resilience Imperative

Of our four community types, residential communities may seem like the least likely to attain the kind of high-level energy resilience described in this book. However, these communities resemble the other campus environments in our analysis in important ways. We describe a community as defined by contiguous land and the institution that occupies it. In a small town, this institution is the local government. Like the other three community types, the *mission* of a local government depends on energy resilience. Again, the safety of constituents is a primary concern. In an emergency situation, the local government is the first and most important safeguard for local residents. In order to perform this function, the municipality must maintain power at fire and police stations to support first responders, emergency shelters to house displaced residents, and emergency operations centers to support and coordinate the emergency response. If these facilities go dark along with the rest of the town, the government will fail in its mission when it is needed most. Recognizing this imperative in the wake of Hurricane Sandy, Governor Chris Christie of New Jersey sought to leverage federal funds to improve energy resilience at the municipal level within New Jersey.[61]

With two $13 million grants from the federal Hazard Mitigation Grant Program (HMGP), the state expects to fund 337 county and municipal energy resilience projects statewide. These funds will be used to support a variety of energy solutions to improve local energy resilience, including microgrids, solar power with battery back-up, and on-site natural gas powered emergency generators. These projects are intended to support life safety facilities like police and fire stations, shelters, emergency operations centers, and lifeline facilities such as water supply and wastewater treatment plants. In a demonstration of the extent of local governments' interest in energy resilience, New Jersey received over $469 million in requests for an initial available funding pool of just $25 million.[61]

Residential Communities—
The Energy Resilience Opportunity

Residential communities have the same characteristics that we have used to identify energy resilience opportunities in our other community types. Municipalities have a core *mission* that depends on power continuous power. They have *place* and *permanence* characteristics that are well suited to energy resilience planning. Municipalities typically possess a fair amount of publically owned land and infrastructure on which new energy projects could be developed. Residential communities may also be the most permanent among our types. As we have noted, it is not uncommon for municipalities to make management plans that extend 50 years into the future. Even with the space and permanence to support large-scale energy-resilience planning, municipalities may sometimes lack the *control* necessary to carry them out. Our analysis of residential communities in this book focuses on public infrastructure—buildings and other assets that are owned by the municipality or community. The kinds of projects funded by HMGP grants in New Jersey will affect public infrastructure and will not require projects or behavior changes from private citizens. However, it should be noted that the support of residents can make a big difference in energy resilience planning at the municipal level.

The level of control wielded by publically elected officials will seldom match that of a central hospital or university administration (or certainly the military leadership on base). Nevertheless, an organized and engaged constituency can make a town supervisor or community administrator a potent agent of change when it comes to energy resilience. The town of Caroline, New York, is a perfect example of this phenomenon. Although the residents of the tiny upstate hamlet are diverse economically and culturally, they share a desire for self-reliance and a deep love for their community. Dedicated organizers were able to harness the power of these shared values to pursue a wide variety of projects aimed at energy independence, from self-sufficient town buildings to a (yet to be completed) project to power the entire town and all of its residents with wind-derived power.

The same fundamental forces were at work when Rockport, Missouri, became the first energy-independent town in the United States in 2008. Like many small Midwest towns, Rockport experienced dwindling fortunes in the age of industrial agriculture. But as in Caroline, a small group of citizens with an idea was able to organize the community around the ideals of self-reliance and resilience from rising energy prices. This group built support for the idea throughout the community and eventually took their proposal to John Deere, which leased a plot of former farm land in town and partnered with St. Louis-based Wind Capital Group to build four wind turbines there.[62] These turbines provide more than enough energy to power the town, and the excess is sold back to the grid. In addition, the wind farm and several others that have sprung up nearby are generating roughly $1.1 million in annual property tax revenue for Atchison country, with Rockport at the center.[63]

These stories offer one path toward municipal energy resilience, but progress can be made without major expenditures or a groundswell of support among constituents for administrators to achieve energy resilience in their community. Although the content of this book will apply to any institution with a physical footprint and central control, we will continue to refer to these four community types for examples and case studies. No matter what kind of community or institution you manage, it is important for every community to benchmark its current position on the energy resilience spectrum, and to plot a course to where it ultimately wants to be on that spectrum.

Part II
Energy Resilience Management

Chapter 3

Institutional Planning for Energy Resilience

Once leaders of an institution recognize the need for increased energy resilience, they are left with the question of how to pursue it. This chapter will explore the energy planning process, from developing a business case for energy resilience efforts to the evaluation and modification of a successful plan. An institutional champion who attempts to move forward with implementing energy solutions without engaging in coalition building and careful planning may have success with one or even two projects, but they will have difficulty in creating lasting, systemic changes that will create a more efficient, more resilient institution. This entire effort is no different from what you would employ in any strategic planning effort, except you are focusing on energy resilience, which imposes some energy-specific considerations as you take the first planning steps.

The energy resilience planning effort should be done carefully. A poorly planned or poorly executed planning effort can derail the entire process or at least mothball it for a considerable period of time. The U.S. Department of Energy's Community Energy Strategic Planning Academy developed a community energy planning cycle (pictured below) that will serve as the basis for this discussion of energy resilience planning. This Academy was offered to local energy planning officials in 2011. One of this book's authors, Brian Levite, was part of the team of energy planning experts who developed this graphic and the course materials for the Academy. This nine-step plan is a valuable roadmap for a planning effort, but it is by no means the only effective approach.

The strategy discussed in this section can be used as a starting point, but the final plan should be tailored to the needs of your institution, the energy challenges you will be facing, and the desires of your stakeholders.

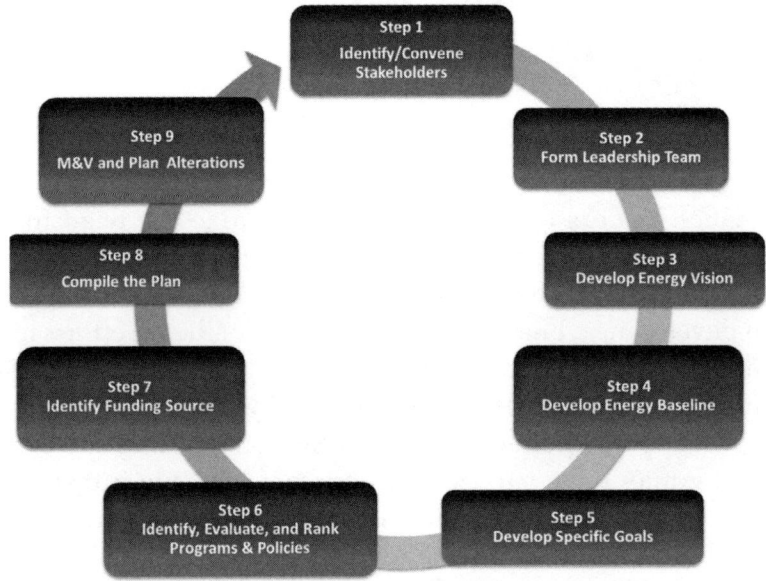

Figure 3-1: The Community Energy Planning Cycle.

Before delving into step one (identifying and convening stakeholders), the champion of energy resilience should build a business case around this topic to justify the planning effort. As with any considerable investment, energy projects require a strong business case—not just to get authorization for funding, but to ensure that all parties involved understand the value of the project and are invested in its success. Note that you do not need to have specific solutions identified ahead of time to address each potential threat to energy security. In fact, allowing other stakeholders to play a role in identifying the specific solutions will go a long way toward developing buy-in and long-term support from these people in the future. Your job at this point is to identify the threats, quantify their impact on your institution, and illustrate a broad vision where the community has solved this issue and is much better off for it.

Energy Resilience Planning

Identifying the overall energy problem in your institution will not be simple. As noted in Chapter 2, energy resilience addresses multiple challenges in a community. The key is to understand which challenge (or set of challenges) is most pressing for your institution and what threat is going to drive sustained action. In addition to developing a business case, you also need to identify the right stakeholders to engage and know how to present the information in a manner that will resonate with them. Finally, understand what funding or financing options are available so you can focus the group's attention on efforts that are possible given budget constraints and contracting avenues. There is not a single path to success, and your approach should be tailored to the needs, interests and strengths of your institution. Below are some hypothetical examples of how different communities might approach this differently.

Rallying around Public Safety

Kate is an urban planner for a town in New York. After Hurricane Sandy, there have been several articles in the local paper and comments at town hall meetings about the fragility of their energy infrastructure. City officials seem responsive to these concerns and want to ensure a safer future. Kate can build a business case around the risk posed to their community by future weather events. She can demonstrate that by making investments in a more resilient community, the town can protect its citizens, avoid the costs of lost productivity and demonstrate forward-thinking leadership in responding to community concerns. Kate reaches out to the mayor, the city council, and the chamber of commerce. She presents a very human-oriented case to show how efforts in energy resilience can make their community safer for those that live there. She also makes the case that this investment will make the community more attractive for businesses or individuals moving to the area. Understanding that local budgets are tight, Kate explores opportunities in energy service performance contracts and other innovative financing approaches so that when someone inevitably says "we can't afford it," Kate can show them ways it can be done with minimal local investment.

Stressing the Green Benefits: Environmental and Financial

Andrew is the head of operations for a college in California. The college has made public pledges to reduce energy usage and improve the sustainability of its operations. The student body also wants a greener campus and has a vocal environmental club. Andrew develops a business case focused on the positive environmental impact that on-site generation, energy efficiency, and even emergency planning can have on the campus. He couples this with analysis of potential cost savings, but he presents those numbers in terms that will resonate with his stakeholders; for example, "...this would allow us to hire [NUMBER] more teaching faculty," or "...this would mean our college could offer [NUMBER] more full-time scholarships to outstanding students each year." Andrew speaks with the school president first and then engages the provost, some environmental faculty members, student organizations (the environmental club and the student government), and possibly even representatives from local businesses or on-site vendors. Andrew can work with the school's finance department to find a way to borrow from future energy budgets to finance projects today.

Using a Mandate to Drive Compliance

Jonathan is the energy manager for a military base in New Mexico. He must follow executive orders from the White House to hit certain energy reduction targets. His base commander has also expressed concern about combat readiness should connection to the local utility fail. Jonathan knows that his efforts cannot be seen as a distraction from the core mission. He also knows he can utilize the leverage provided by executive orders and a risk-averse culture to make changes. As he discusses the business case, he will focus on what might happen to the base and its combat readiness should they experience widespread power loss or noncompliance with an executive order. Jonathan engages commanders on base, the local community's planning or sustainability office, and businesses that operate on base. He can work with the Office of the Assistant Secretary of Defense for Operational Energy to obtain financing to support these kinds of measures.

In each of these cases, the champion of energy resilience had to understand the drivers for their institution, the metrics by which that institution will measure needs, and the stakeholders that must buy into this concept before progress can be made. In his book, *The Industrial Energy Harvest,* Christopher Russell writes that quantifying the cost of doing nothing is the critical element to making the business case for energy efficiency efforts.[64] Typically, those costs are focused on energy expenditures that can be avoided. In the case of energy resiliency, we can add those bills to the potentially catastrophic costs of losing access to energy for an extended period of time.

When calculating the financial cost of not pursuing energy performance projects, the approach can be quite simple. When evaluating investments with metrics like net-present-value (NPV), return on investment (ROI), internal rate of return (IRR) or simple payback, you can leverage online calculator tools from organizations like the Department of Energy,[65] the Environmental Protection Agency,[66] and a wide variety of non-profits, utilities, and product manufacturers. Using tools like these is highly recommended to calculate the potential energy saved, greenhouse gases avoided, ROI, NPV and IRR of potential projects.

Analyzing these numbers is just the initial phase of building the business case. To create a truly compelling case, these numbers need to be presented in metrics that will resonate with your key stakeholders. Examine the metrics of success or progress used by the organization to evaluate success in their core mission and attempt to translate the benefits of energy programs into those metrics. That could involve citing measurements such as "equivalent to planting [NUMBER] new trees," "could provide full scholarships for [NUMBER] new students each year," or "equivalent to [NUMBER] new M1114 Up-Armored Humvees for our battalion."

When calculating the toll that an energy outage can have

on an institution, the approach is not quite so clear. This cost is based on an estimate of the risk of a disaster (e.g., a weather event, earthquake, accident, terrorist attack) and the risk that such a disaster would disrupt the supply of energy (electricity, natural gas, heating oil, gasoline or a combination of all of those). Historical data on these events may not be a reliable source as shifts in climate, infrastructure stability and the geopolitical situation make the past an unreliable predictor of future events. If we could calculate the likelihood of all of these factors, we would then need to determine how long those energy services were likely to be out and, finally, what specific impacts that might have on the institution. It's even more complicated based on the fact that even if gasoline is no longer available in one community, the impact of scarcity can be mitigated by an available supply in the next town. If natural gas and heating oil become unavailable, the impact of this situation would be radically different given the season. The same can be said for electricity.

This is not to say that you shouldn't try to estimate the impact of potential energy interruptions or the value of energy resilience. However, making accurate quantitative estimates about the likelihood and costs of energy failure is exceedingly difficult. Instead, we propose that communities present possible and even likely scenarios and scope out the costs should those come to pass. This kind of scenario-based evaluation of potential risk is not as scientific, but it is a valid starting point for a policy discussion. Alternately, if your community has, in the recent past, sustained some kind of energy-related challenge, you can use data from that event to present a plausible scenario. In the end, you must acknowledge that each stakeholder will have a very subjective reaction to this argument, judging the likelihood of these scenarios and how pressing the threat is to your institution.

The first thing to make clear about the potential threat to the energy system is that you are not talking about short, manageable power outages—the kind most people have experienced in the past in which people unplug from the world, take a breath, and then go back about their business in an hour or two. The scenario we are presenting is the threat of long-term energy loss—what parts of Japan suffered after the tsunami of 2011, and what the New York City area saw after Hurricane Sandy. In both cases, communities were left attempting to deal with situations that had not been anticipated: weeks without electricity supply, compounded by physical damage to infrastructure. Residents were left in the dark and the cold. Businesses were shuttered and many went out of business permanently, unable to cope with the lost revenue. Public safety was compromised. While insurance can replace damaged buildings and equipment, it does not cover lost revenue from having a business closed. It is important to find one or several scenarios like this that your stakeholders will see as potentially relevant to your institution. On the East Coast of the United States, hurricanes may seem most likely. In the middle of the country, tornados may be the concern. On the West Coast, earthquakes are a clear threat. Our aging infrastructure and the treat of terrorist attack are viable concerns in any part of the country.

Although this book has focused mainly on addressing energy outages, disruption in power is not the only energy-related threats Energy price spikes or even steady increases represent another threat. And, while you can make the case that price increases are more threatening to budgets than people, price volatility remains a key issue for business and community leaders. If, over the course of a year or two, energy costs rose from 3 percent of your institution's operating budget to 7 percent of that budget, the difference would represent a serious threat to financial health. Energy resilience efforts can address these issues along with the challenge of supply interruption. Evaluating which of these is the larger perceived threat will definitely have an impact on your planning process.

Once you have identified one or more scenarios, consider how the realization of those scenarios would impact your institution. What resources are in place to address those impacts, and how would the members of the community, businesses, and key operations be affected? [For a more in-depth discussion of how blackouts can affect a community, read CRO Forum's November 2011 position paper titled "Power Blackout Risks: Risk Management Options."] Illustrate these concerns in short case study examples for stakeholders to consider. The goal here is to avoid fear mongering to support your cause. Rather, encourage your partners to think honestly about how some entirely possible risks might affect the institution and make their own judgments on the value of taking precautionary measures. Pursuing a more energy resilient institution is in many ways akin to beginning a diet and exercise program. In this case, pursuing energy resilience can help avoid catastrophe, but it also has very real, tangible benefits that can benefit the institution even if an energy crisis never arises.

STEP ONE: IDENTIFY/CONVENE STAKEHOLDERS

Once you have completed your analysis and can effectively convey the business case of your energy resilience projects, the next step is to present your case to colleagues and stakeholders to build a coalition of support and convene a planning session—two distinct but equally important tasks t. As a preliminary step, identify the stakeholders, the groups and individuals that would influence an energy resilience project. In some cases, you might want to build support for the idea in general, using meetings with stakeholders and your customized business case information to discuss the importance of energy resilience. In other organizations, the best approach might be more direct: getting the right players in the same room to determine the importance of energy resilience and outline parameters for developing an energy resilience plan.

The Department of Defense's Net-Zero Military Installations planning guide[48] offers a valuable list of participants for this kind of effort.

- Installation management (such as the base commander)
- Energy manager
- Facilities and maintenance personnel
- Fleet vehicle manager
- Director of public works
- Contracting officer

Energy Resilience Planning

- Environmental manager
- Master planning
- Installation public affairs officer
- Installation security officer
- Utility company

For other types of communities, some different players may be involved. For instance, at a hospital, they may include:

- Hospital administrators
- Energy management and maintenance staff
- Potential business partners
- Doctors and patients' representatives
- Union representatives
- Utility representatives

Ideally, you will have developed some information (in writing or in the form of a verbal elevator pitch) about the business case for addressing this issue. You may have even developed this business case using metrics important to your organization. At this point, you can consider customizing this business case even further to reach potentially resistant audiences. For example, when speaking with local businesses, you could talk about the impact other natural disasters have had on local businesses and demonstrate the value of this engagement to protecting their interests. Some research on actions other communities have taken in this area can be valuable in demonstrating that these concepts are already being applied.

Facts are powerful, but you should also realize that you are selling a new concept, and some good old-fashioned flattery can bring stakeholders to the table. Reassure them that they are important players in this area and that their participation is critical to making the effort a success. Tell them their specialized skills and insights are critical if this effort is going to be properly evaluated (a statement which is likely to be quite true). Each of the stakeholders you will be attempting to engage will contribute a specific set of competencies, and bring their unique perspective to bear.

Rather than simply recruiting as many participants as possible, direct your efforts to those parties who are most likely to move the process forward or help you identify potential roadblocks. An energy resilience planning effort can suffer from too much involvement as well as too little. Whether you are attempting to build support for the concept in general or convening a strategic planning meeting, focus your efforts on people and organizations that 1) you will need to support the concept from a political or funding perspective 2) you will need to implement the project ideas that come out of a planning effort or 3) could apply concepts in their own portfolios to complement your efforts. Recruiting people who will not have a substantive role in energy resilience efforts can be detrimental since you want to facilitate in-depth discussion from critical stakeholders. An energy planning effort should probably involve between 6 and 12 stakeholders.

In the case of a municipal government, you will most likely want to involve representatives from the community. This can be done at regularly scheduled town hall discussions, but some special planning and advance education will be critical here. Your citizens are likely to be less familiar with aspects of energy management and generation. Some may come to the discussion with misconceptions. It is highly recommended that you begin any public discussion of these issues with a brief presentation on the threats to energy resilience, economic analysis on investing in energy resilience projects, and even some of the technical aspects of things like energy efficiency, energy storage and on-site generation.

If you do not have a mandate from senior leadership to move forward with an energy resilience planning effort, you may need to convene stakeholders to make the business case and garner feedback on how to proceed. The first goal of this meeting is to understand the opinions of each stakeholder on the importance of energy resilience and what they would be willing to contribute to efforts on that front. The second goal is to get all parties to agree on a path forward. This is critical. Do not let the stakeholders leave the meeting until there is general agreement on what will be done next, who will do it, and when it will be done. You can try to begin developing an energy vision in this meeting, but creating a vision statement in a committee may be difficult. You may be better served by discussing the motivations and priorities of the stakeholders involved.

As a best practice, take good minutes at this meeting. The following are some examples of the notes that would help guide next steps:

- The stakeholders generally agreed that energy resilience was an important initiative to focus on.
- There was some dissention from the local utility, which felt that distributed generation projects could be problematic for their generation planning efforts.
- Several stakeholders agreed that, while this is a priority, pursuing these efforts should not distract from [*insert other project priority here*].
- The Energy Director has been tasked with developing an energy vision statement and identifying energy performance opportunities. He will convene a meeting by the end of October with a subset of stakeholders to determine a technical approach here.
- The Mayor/Commanding Officer/Administrator/President and Budget Director will have an offline discussion to attempt to identify additional funds that can be used in this fiscal year to support the Energy Director's analysis. They will provide guidance by October 15.

Several of the steps outlined here will require meetings of your stakeholders. The best way to ensure that these meetings are successful is to identify a qualified facilitator to design and run them. Most professionals attend hundreds of meetings of varying value each year. The way to make the best use of your stakeholders' time and generate the outcome you are looking for is to have one person whose job it is to design and run the meeting. This person should not have much to say about the content of the meeting because participation in that sense undercuts their ability to effectively facilitate. If you do not have access to a trained facilitator and have no budget to hire one, try to identify someone who is well liked throughout your organization or who is known in the community. Again, preferably, this will not be someone who will play an important role in deploying energy resilience projects. Select someone seasoned enough to be taken seriously, but not so senior that they will intimidate participants or feel the need to make their opinions known during the meeting. The meeting facilitator's goal is achieve the meeting objectives and ensure that all stakeholders have a chance to contribute to the thought process.

The case study below outlines the planning effort undertaken by Fort Bliss in El Paso, Texas. U.S. military bases have been among the most aggressive communities in pursuit of energy resilience. Although we may think of the U.S. military as an extremely hierarchical organization where priorities are identified and then executed, in truth, military groups can respond to official directives in a variety of ways. Determining the correct course of action requires discussion, expert advice, and business case development. Fort Bliss is an excellent example of how an energy resilience effort was carefully considered and deployed in partnership with numerous stakeholders.

> **Case Study: Fort Bliss, El Paso, Texas[67]**
>
> Energy resilience efforts at Fort Bliss serve as an excellent example of leveraging stakeholder engagement to achieve energy and water goals. The process began with directives from Washington that the military should reduce energy consumption. This call was taken up by Fort Bliss Commanding General Howard B. Bromburg, who felt that conservation and efficiency were important values to champion. His successor, General Dana Pittard, championed the same priorities after Bromburg's departure. However, the enthusiasm of these two leaders was not enough to realize the plan.
>
> Staff like Energy Branch Chief BJ Tomlinson had to engage a wide variety of stakeholders in order to achieve the goals set forth by the

base commanders. Tomlinson's organizing activities included:

- Partnering with the newly created Army Net-Zero Initiative and volunteering Fort Bliss to be a pilot installation to receiving technical support on program development;

- Working closely with local governments—Fort Bliss has a footprint that spans both Texas and New Mexico. As new projects were considered (particularly new power generation projects), the local governments and utilities were at the table creating a project plan that was valuable to all involved. These interactions were not always smooth, but these relationships have improved over time and Chief Tomlinson believes that finding common ground on new projects is getting easier.

- Working with energy performance contractors to complete efficiency projects when additional capital funding for efficiency projects was not available.

- Enlisting the support of the base commander to circumvent red tape and fast-track projects with clear energy resilience and efficiency benefits.

These collaborations have delivered real value to Fort Bliss. The base has implemented a wide variety of energy saving measures in their 2,200 buildings, encompassing 32 million square feet and representing a $357 million annual energy buy. Many of these measures were focused on load shifting as peak demand costs are a huge expense for the base. A 24 megawatt on-site generator has been installed, although it will require further upgrades in order to allow full control over load selection. In addition, expanded bike trails and bus service has made the base more accessible and reduced car emissions.

Base leadership understands that until they have energy storage, more on-site generation, and a working microgrid, risks to external energy supply will mean risks to the operational readiness of Fort Bliss. More collaborations are planned for the future. The base and local government are working together to connect bike lanes and bus services to further facilitate non-car travel across the base. To better manage waste, the city is interested in working with the base to develop a joint waste-to-energy plant of

> 40-45 megawatts.
>
> Fort Bliss is dramatically reducing its energy intensity per square foot. They are on track to meet the federally mandated energy savings goals by 2015 and exceed them after that. They are beating their own goals for water conservation. The reason for this success is clear: open communication and collaboration of diverse stakeholders toward common goals.

STEP TWO: FORM LEADERSHIP TEAM

The leadership team will be a small group that will be responsible for moving the process forward and executing most of the plans developed by stakeholders. Typically, this group should be small (3-5 people) and include the energy director or whomever in the organization leads energy work. It is critical to have a small group sharing the workload as opposed to relying exclusively on the energy director to lead the effort and to engage different groups as the plan develops. The members of the leadership team, serving as the project owners, will need to meet regularly to make sure this project does not quietly die off.

In terms of composition, attempt to represent each of the major groups that will be involved with implementation of any plan you develop. Unlike with your stakeholder group (where you are engaging a broader audience), this team should be comprised entirely of people with decision-making authority within the organization. Including the right players early on will be critical for several reasons:

- Ensuring a variety of viewpoints
- Improving visibility around potential roadblocks
- Securing buy-in to generate continued support
- Creating more linkages to stakeholder groups for project planning
- Distributing the workload so the energy champion does not become overwhelmed.

ENERGY RESILIENCE PLANNING

STEP THREE: DEVELOP AN ENERGY VISION

A planning effort cannot be successful without a clear vision. As we have discussed, energy resilience can have a variety of drivers and can manifest itself in different ways. Is your institution looking to become fully energy independent? Is it just looking for protection from price shocks? Are natural disasters the major concern? Working with the stakeholder team to understand the drivers of your institution's efforts will allow you to develop a strong foundation for a concrete energy vision. The vision needs to be realistic and functional. It should be flexible enough to allow the planning team to evaluate multiple approaches. When developing language for this vision, imagine how a group of strangers would develop your actual energy resilience plan based only on the vision language. Would they know what you wanted to do and when you wanted it done? See Table 3-1.

STEP FOUR: DEVELOP AN ENERGY RESILIENCE BASELINE

The adage "you can't manage what you don't measure" is so fundamental to sound energy resilience it has become something of a cliché. An energy resilience baseline effort determines the current state of your technology, infrastructure, skill sets, and operating procedures in each area of energy resilience. The next chapter in this book provides a maturity model template for your organization to use in understanding the state of your energy resilience approach now. There is more in-depth discussion of how to conduct benchmarking and analysis on the three critical areas of energy resilience (energy management, on-site generation, and emergency planning) in Chapters 5, 6 and 7.

STEP FIVE: DEVELOP SPECIFIC GOALS

Once an energy baseline is completed, you can set specific goals around energy resilience. It is important to have active stakeholder involvement in this stage and an in-person, facilitated goal-setting meeting. Remember that this meeting should not discuss whether or not energy resilience is important—that should

Table 3-1. Energy Vision Statements

Example Vision Statement	Evaluation
Lilliana Township will utilize cutting edge technologies to become a net-zero community and the most energy resilient town in the United States.	This is too ambitious. The planning team will likely soon find that achieving this vision is impossible, leaving them with no functional vision.
Greenwood College will evaluate energy resilience opportunities and attempt to reduce energy consumption by 10 percent.	This is not ambitious enough. It focuses on process over results, and is prescriptive in terms of what will be addressed. If energy efficiency is your goal, that 10 percent figure may make sense (though it is very modest). If energy resilience is your goal, efficiency may or may not be part of the strategy.
Hobart Medical Center will employ highly efficient cooling and lighting technology combined with a micoturbine generator to allow 80% of our facilities to operate in island mode when necessary.	This is too tactical. It ties the vision to specific technologies. It sets specific technical goals that are likely unknowable at this time. *Setting specific goals will be covered in Step 5 later in this chapter.*
By 2020, Fort Birch will be capable of supplying energy to 85% of its operations even when power is not available from off-site sources.	This is a good energy vision statement. It is ambitious but realistic. It includes a timeframe. It is clearly focused on reliability.
The Maerin corporate campus will leverage advanced technologies to reduce energy costs by 70% by 2030, demonstrating our role as a global leader in social innovation.	This is also a good energy vision. It is clearly focused on risks associated with energy costs but it also ties this effort directly to the corporate mission.

have been established in Step 3. The purpose of this meeting is to identify the goals that will help the institution achieve its energy vision. The output can be a final set of agreed goals or a series of goal recommendations for the ultimate decision maker to address (depending on your process). It is not important to work out technical details of every tactic in the plan. Instead, focus on capturing and understanding the priorities of the stakeholders. If the stakeholder meeting is just about providing input, the leadership team or organizational leader will select a final set of goals from those generated in the meeting. Figure 3-2 and Table 3-2 give some examples of energy resilience goals. In each case, the first example is more output-based while the second is more outcome-based. Both are acceptable at this stage. In practice, each of these goals would have a specific time-frame attached to it.

STEP SIX: EVALUATE AND RANK PROGRAMS

Once a set of goals is finalized, the leadership team should develop a set of tactics to achieve those goals. Ideally, a diverse set of tactics will be developed so they can be evaluated, and the most promising can be selected. While any analysis is likely to identify the cost effectiveness of a tactic, it is important to also focus on how well each tactic will achieve the specific goals outlined. Here are some examples of specific tactics you might consider and what metrics you could use to evaluate them. See Table 3-3.

At this point in the process, you will have developed an overall energy resilience vision, goals to help you achieve that vision, and tactics you will employ to achieve those goals. The framework should look something like the image below. Some members of your stakeholder engagement group may want to switch right to tactic development. You may hear statements like "Why are we worrying about goals when we should just figure out how to build this new CHP project?" It is critical to have a structure of vision, goals and tactics, however, to ensure that a holistic effort is designed so that each project developed supports the greater effort. Having a vision and goals also ensures that all participants can evaluate tactics using the same parameters for success.

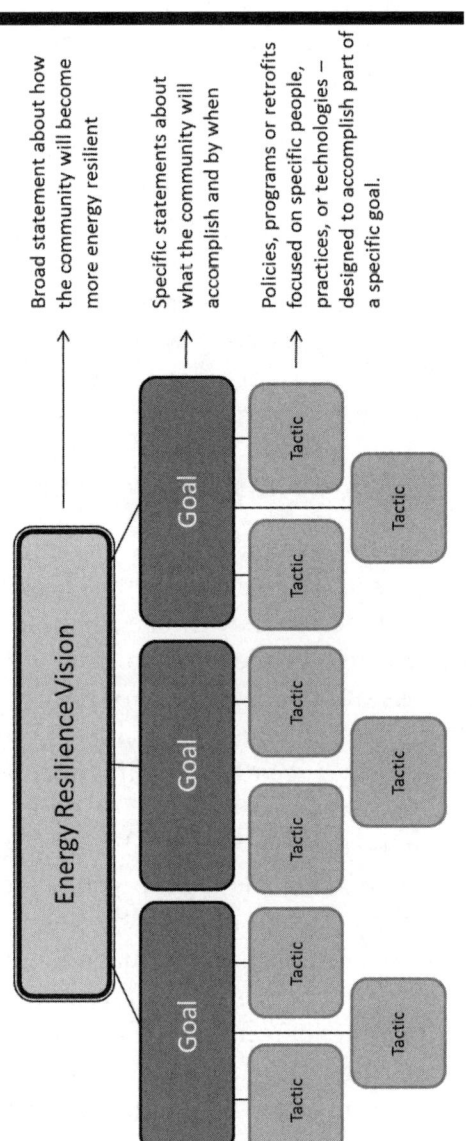

Figure 3-2: Energy Resilience Plan Structure

Table 3-2. Specific Goals

Type of Goal	Examples
Energy Purchasing	• Negotiate energy purchasing contracts so that 75% of the energy purchases for the next 10 years are at a fixed, predictable rate. • Reduce our cost of purchased BTU by 15% by [DATE].
Energy Efficiency	• Fund any identified energy conservation measure with a payback of three years or less. • Reduce building energy consumption by 20% within five years (using the previous calendar year as a baseline)
Transportation *(more applicable to municipal governments)*	• Improve deployment of bus routes by [DATE] so that no member of the community would need to walk more than one mile to reach a bus heading downtown. • Reduce average vehicle miles traveled for our constituents by 10% in the next five years.
On-site Generation	• Develop on-site generation that can handle 50% of our institution's electricity load for at least three weeks in the case of a grid failure by [DATE]. • Eliminate our reliance on delivered electricity within ten years.
Renewable Energy	• Deploy 20 megawatts of renewable energy generation by [YEAR]. • Secure 30% of our energy from on-site renewable energy sources by [YEAR].
Microgrid	• Deploy a power control system and switches that will allow us to power [LIST OF 2-4 CRITICAL FACILITIES] with on-site energy sources by [DATE]. • By [DATE], deploy a microgrid that will allow our institution to operate in "Island Mode," providing what energy we generate on-site to the highest priority loads (as outlined in the Emergency Response Plan).
Emergency Planning	• Educate our community on how the energy systems will respond in the case of various kinds of power outage events so they can plan accordingly. • Integrate emergency response and power control activities so we can continue delivering critical services within 30 minutes of a blackout event.

Table 3-3. Evaluation of Programs

Vision	Example Goal	Tactic	Evaluation metrics
Energy independence	Develop on-site generation to handle 50% of our institution's electricity load for at least three weeks in the case of a grid failure by [DATE].	Build a 25-35MW gas-fired cogeneration plant next to the existing central plant.	• Cost to develop the plant • Anticipated payback based on expected gas rates and electric utility rates • Percentage of the institution's load that can be handled • Security of the natural gas supply
Insulation from price spikes	Negotiate energy purchasing contracts so that 75% of our energy purchases for the next 10 years are at a fixed, predictable rate.	Secure bids from third party energy suppliers to evaluate savings options.	• Likelihood of electricity prices rising or falling [The most reliable source for this kind of information is the US Department of Energy's Energy Information Administration: http://www.eia.gov/] • The level of rates that our organization is able to negotiate and how this compares to current energy rates • Potential cost savings compared to another approach (such as developing our own on-site generation)
Environmental sustainability	Reduce overall greenhouse gas footprint by 20% in the next fifteen years.	Leverage energy performance contracting to reduce energy consumption in buildings 20%	• Possibility of a 20% reduction given past efforts and current technology • Ability to use internal capital to finance energy performance improvements and thereby retain more of the savings? • Planned changes to the building portfolio that might affect this effort

STEP SEVEN: EXPLORE FUNDING SOURCES

Although there are going to be a few low-cost ideas you can execute in an afternoon, most of the tactics you will develop are going to require investment. At this point, the leadership team needs to determine how each of these tactics is going to be funded. There may be one funding mechanism chosen to cover all tactics, but it may be valuable to think about them individually to understand the best financing approach for each. In many institutions, financing projects can be an upfront sticking point; you may need to give some thought to this step early in the process. Still, the wide variety of financing mechanisms available today means any institution can find some funding options to make progress on energy resilience. Here are several approaches:

Internal Capital

For certain projects, the institution may be able to dedicate part of its annual budget to pay for projects.

Pros: Traditional approach ensures the institution reaps 100 percent of the rewards from their efforts.
This can be a good route for projects that create energy security but it may be much harder to secure external capital for projects that may not produce energy savings.

Cons: Getting large allocations from annual budgets is typically very difficult. Using the annual budget process may delay action.

Borrowing from Energy Savings

For projects that are projected to reduce energy costs, the finance department may allow for money to be borrowed from those future savings in order to pay for projects that generate them. Municipal governments can also use bond measures to employ this approach.

Pros: Creates a budget-neutral approach that still keeps all savings in-house.

Cons: Generates risk in that, should projects fail to secure anticipated savings, energy budgets will fall short. This can only be used for tactics focused on reducing energy consumption.

Grant Funding

The institution works with government agencies or non-profit organizations to secure program funding or grant funding to pursue these projects.

Pros: Can be used to pursue any kind of project, including projects that do not yield actual dollar savings.

Cons: Requires an investment of staff time to pursue grants or Federal program dollars. Also, grants often only cover part of an effort.

Leveraging Private Investment

The institution can work with private businesses (such as vendors, tenants, landlords, utilities, etc.) to encourage activities that support the energy resilience plan. This may simply involve convincing the businesses of the value of these activities. Municipal governments can use things like preferential zoning. Companies, universities and military bases can negotiate these activities into vendor contracts.

Pros: Generates action on energy resilience without requiring internal expenditures.

Cons: Programs created this way are largely out of the control of the leadership team. Performance toward goals may suffer in cases where partners deploying the tactics are simply doing it to satisfy a requirement.

Energy Service Performance Contracts

Private engineering firms conduct energy audits of a customer's facilities and identify energy savings opportunities. The firm then performs the identified retrofits using their own money and the client pays for this service by giving the firm a portion of the energy cost savings over a pre-determined period.

Pros: Creates an easy, low-risk avenue to achieve energy savings without waiting on internal funding.

Cons: The institution will not get the full benefit of the energy savings. This can only be used for tactics focused on reducing energy consumption. The value of these projects appears as debt on the balance sheet.

Saved Power Purchase Agreement

In this relatively new model, a third-party firm replaces inefficient energy-using equipment with their own, more efficient equipment at no cost to the owner. This retrofit typically includes monitoring capacity that allows the firm to track actual energy saved by the new equipment and compare it to a baseline of energy usage by that equipment prior to the retrofit. The institution agrees to buy a certain number of avoided kilowatt hours (based off that pre-retrofit baseline) at a set price (typically lower than the current utility rates). If no energy is saved, no payments are made. Once all the agreed to saved kilowatt hours have been purchased, the firm abandons their technology in place.

Pros: This approach creates an easy, no-risk avenue to achieve energy savings without waiting on internal funding. The institution is simply purchasing a service so the cost of the project does not appear as debt on the balance sheet.

Cons: The institution will not get the full benefit of the energy savings, as some profit must be made by the third-party firm. Due to the complexity of the approach, this kind of project typically only works well for organizations with a large energy spend, where the investment in contract development and performance monitoring equipment will be justified.

Revolving Loan Fund

In this model the institution allocates money into a fund that is then used to pay for energy performance projects. The savings from those projects go back into the fund and can be used to support other projects in the future. The goal is to create a virtuous cycle where savings from past projects will fund the capital costs of future projects. The use of these revenue sources must be strictly dedicated to energy projects, so that other priorities within the institution are not allowed to drain this fund, and defeat this approach.

Pros: Revolving loan funds have proven to be one of the most successful approaches for supporting continuous energy savings programs. This requires no investment after the initial seed funding.

Cons: This model requires a sizable up-front investment. Projects must be well evaluated and monitored to ensure savings are realized in order for the fund to properly sustain itself. This can only be used to fund energy saving measures.

Figure 3-3 from Jen Weiss' article "Revolving Credit—All Grown Up"[68] shows how a revolving loan fund works:

Third-party Incentives

Whatever funding source is leveraged, look at every source of financial or service support that exists. Many utilities have goals around demand reduction and provide program or financial support to their customers on this front. There are incentives and rebates available from both the Federal and state governments for a variety of efficiency and on-site generation projects.

There are a wide variety of financing options, and the list above likely does not cover them all. When evaluating financing approaches, consider the following questions:

1. Will we be able to pursue this kind of funding? Will leadership approve it? Are there any legal issues?

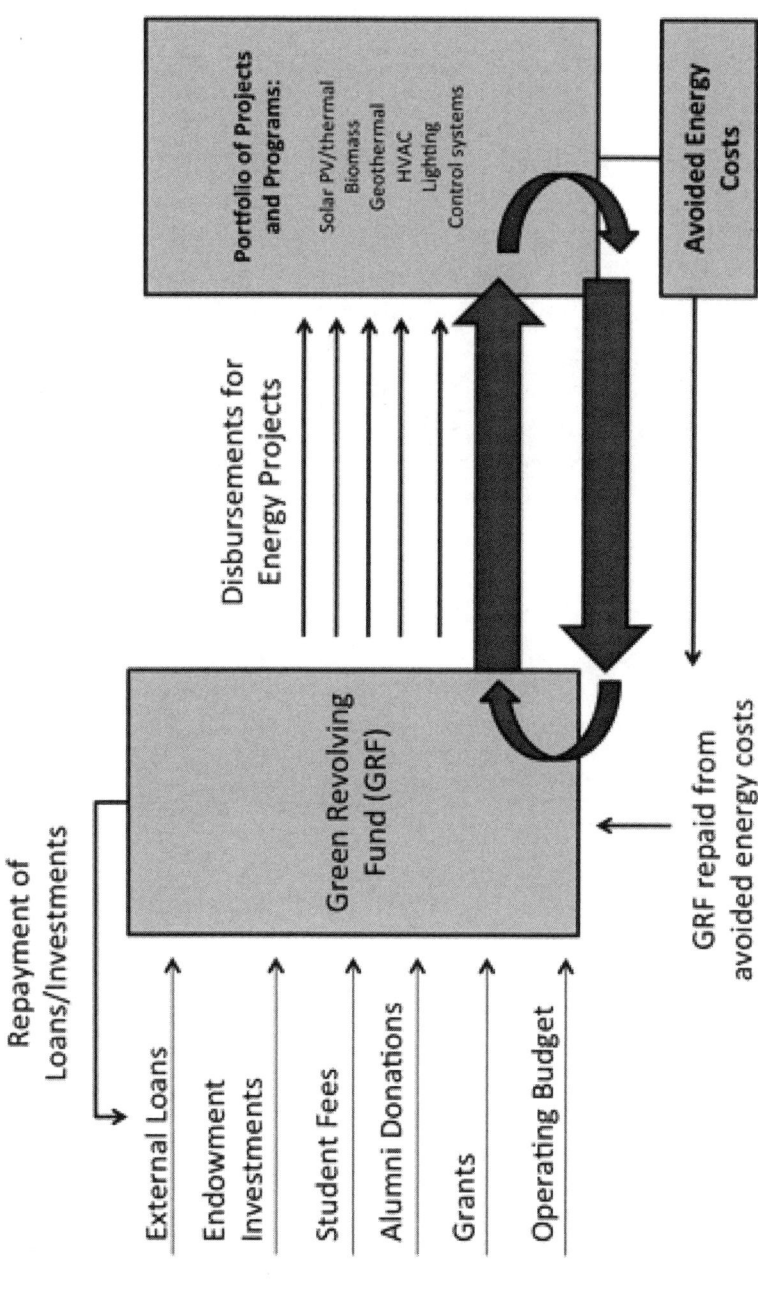

Figure 3-3: Green Revolving Fund[68]

2. Will using this kind of funding impact the effectiveness of the energy program in achieving stated goals?

3. Will this be a sustainable approach? Will using this mechanism help or hinder our efforts to do similar work in the future?

Whatever financing approach you select, consider how that mechanism can be used to generate the maximum possible impact on the energy resilience of your institution. One best practice to consider is bundling measures into larger projects. Projects with higher internal rates of return (faster payback) can help to offset the lower IRR of projects that may have other valuable benefits when it comes to energy resilience. For example, if your institution requires 20 percent rate of return for capital investment, you could pair a lighting upgrade with a 50 percent internal rate of return with a power storage project with a 10 percent rate of return. Bundling projects also helps when you are dealing with projects that have IRR that is harder to define or that involves more uncertainty. It's about creating the largest possible portfolio of activities within your available financing mechanism—mixing rates or return, ancillary benefits and levels of uncertainty.

STEP EIGHT: COMPILE THE PLAN

Once your team develops your energy resilience plan and a financing strategy, publish the plan so it can be easily shared with parties both within and outside of your institution. This may seem like a step that doesn't warrant much explanation, but there are few considerations that will help ensure that the plan is understood and received well by your target audience.

First: Be sure to write up the plan as simply and clearly as possible. Imagine that your entire team wins the lottery and departs for your newly acquired private islands. The team that will replace you needs to understand the energy vision for your institution and understand how those tactics are to be deployed in support of that vision. You should also write this plan for a non-technical audience, to the extent possible. This plan is to be your constituents' doorway into what is being done to ensure the energy resiliency of their town, their base, their campus, their hospital.

Second: Get this plan officially "signed" by the senior official in your institution and have that person share this plan with the entire community. This leader's seal of approval will provide the leadership team with leverage as they attempt to implement the plan. Also, letting the entire community know about the plan will foster good will toward your effort, prepare them for changes they may see as a result of it, and may even generate some new ideas or new allies as your vision, goals and tactics are shared with a wider audience.

Finally: Work with the leadership team (and possibly the stakeholders) to agree on a date to evaluate results. Also designate one person to be responsible with convening the group for that discussion.

STEP NINE: MEASUREMENT, VERIFICATION AND PLAN ALTERATIONS

Any energy resilience plan represents the best thinking of your team at that time. However, times change and new information allows us to refine our approach and more expertly pursue our goals. Your leadership team should be evaluating the progress of each project initiated by the plan. Is it progressing at the expected pace? Are the results expected being achieved? Are those results driving the goals as expected? The worst thing an institution can do is develop a plan that then goes into a drawer and is never heard of again. The second worst thing an institution can do is stick rigidly to elements of a plan that are proving to be less than effective in achieving your goals.

The easiest way to accomplish this is to simply build measurement, verification, and a formal review of your energy resilience efforts into the plan itself. The lower level tactics should be reviewed regularly while the higher level strategies and goals should be reviewed only after some time has passed to provide insight as to the results of the plan. Specific tactics and projects should likely be reviewed at least annually to ensure that they are on track and yielding the intended results. Goals may be addressed annually or every few years. The energy vision should hopefully remain the same until it is achieved, but there may be cases when the leadership team feels the need to alter course and a revisiting of the energy vision is helpful.

Making Your Planning Effort a Success

As you have seen in this chapter, effective energy resilience planning is not that different from other community planning projects. It involves convening critical stakeholders, developing a tiered approach toward a single vision, and leveraging your institution's strengths to make progress and build upon that momentum. Energy systems are complex and interdependent, which is a major reason why they sometimes fail and why that failure can be such a problem. The key is to attack the problem systemically and leverage advances in one area to support advances in other areas. Consider, for example, the way better lighting can lower cooling costs or the way an energy storage project can help make solar power more cost effective. If your organization can successfully engage energy systems as a whole, energy resilience efforts will yield better, more immediate results.

The work done in Caroline, New York, is an excellent example of how a wide variety of stakeholders were brought together to develop a vision, set clear goals against that vision, and undertake specific activities to achieve those goals.

Case Study: Caroline, New York[69]

Situated among the picturesque hills of New York's Finger Lakes Region, the town of Caroline is home to an unusually high degree of political and socioeconomic diversity. A former farming community, Caroline's roughly 3,000 residents are a mix of largely conservative life-long residents and people who settled to work in neighboring Ithaca and Cornell University. Instead of allowing these political differences to create social rifts, the residents of Caroline have come together around the values that unite them: a strong independent streak, and an affinity for self-reliance. Dominic Frongillo, the town's Deputy Supervisor until 2013, says it is these values that have propelled Caroline to a series of impressive victories in the areas of energy conservation, renewable generation, and energy independence.

It all started in 2004 when a contingent of town board members, concerned about inaction on climate change at the national level, came together to donate a portion of their salaries to the purchase of renewable energy for municipal uses. The example they set was infectious, and led to a community fund raising effort that generated enough money to meet all of the town's municipal power needs with wind power for three years, making Caroline only the second municipality in the state to meet all of its public energy needs with renewables. Town residents, however, were not content to rest on their laurels. As Frongilo explains it, "a gathering of citizens was called to address the question, 'why are we paying a premium on wind power owned by a foreign company?' What would it take to create renewable energy owned by the community, and bring the power back home?"

This turned out to be the first meeting of what would become "Energy Independent Caroline" (EIC), an organization of town residents and council members dedicated to increasing Caroline's independence from fossil fuels. The group decided that their strategy would need three fronts: reducing municipal energy use, reducing residential energy use, and the group's white whale—a local renewable generation project equal to the slimmed-down demand of the town as a whole.

The group performed energy audits on all public buildings, and performed energy retrofits on all municipal buildings to maximize efficiency. This effort coincided with the design and construction of a new municipal office building adjacent to the historic town hall. The new building featured a suite of energy efficient design elements, such as solar tubes for daylighting, and took advantage of geothermal heat pumps for heating and cooling, and photovoltaics for on-site energy generation. Taken together, the design of the new building and upgrades to the historic town hall allowed both buildings to meet all of their energy needs with on-site generation. It is worth mentioning that these upgrades would not have been possible without the financial support of Caroline residents, over 80 of whom pitched in at the time to help fund energy projects for public buildings.

Frongillo notes that once the community was organized around the goal of energy independence, they stayed quite active. In pursuit of the group's second goal, to reduce overall energy demand in the town of Caroline, EIC members teamed with Cornell Cooperative Extension to distribute information on energy conservation. In the fall of 2008, over 100 volunteers went door to door delivering CFL bulbs along with information on reducing home energy use to Caroline residents. By the time the day was done, they had reached nearly every one of the town's 1,400 homes. EIC members calculated that if every resident used the bulb that was handed out, the community would save $70,000 over the life of the bulbs.

While pursuing the group's final goal of local energy generation, Caroline residents researched several options, and convened a community conversation about community-owned wind power. After preliminary studies in collaboration with students and faculty at Cornell University, residents realized that it might be more cost effective to share resources with wind energy advocates in neighboring town, Enfield, who are was pursuing the development of the planned 11.9 megawatt Black Oak Wind Farm. Caroline residents actively organized events to encourage residents to buy shares in Black Oak. When the project is completed in 2015, it will become the first community-owned wind farm in New York State and residents of Caroline will succeed in bringing clean energy to their community.

Energy Independent Caroline has served as a model for other communities and has driven several regional initiatives, including Solarize Tompkins Southeast. Spearheaded by residents in Caroline, Dryden, and Danby, the pilot solar purchasing program led in 2014 to a major new program, Solar Tompkins, that is bringing affordable solar power to 1,310 rooftops in Tompkins County—and will fully double the county's solar energy production in just one year.

Although the regional power grid will be part of the energy infrastructure at Caroline for the foreseeable future, all of the groundwork is now laid for greater energy resilience should it be pursued by Caroline residents in the future. Public facilities have been made more efficient, private homes are slimming down their demand, and a large scale local generation project is underway. These steps can certainly serve as a positive example for other small communities that are seeking energy resilience. Perhaps more important than the details themselves, however, is the lesson from how these steps were accomplished in Caroline. Each victory has been a testament to the effective community organizing that is necessary to bring a diverse town together around the goal of energy independence. As Frongillo explains, "The name Energy Independent Caroline represents the kind energy transition that we can all get behind. The concept of producing clean energy in our own community so we can keep our money in the pockets of families instead of sending it oversees to giant foreign corporations, that's something that's easily supported by conservatives, progressives, republicans, democrats—everyone. We are doing our part to fight climate change, but we are also honoring our rural ethic of independence and our values of stewarding our own resources and being the masters of our own destiny."

Chapter 4

Energy Resilience Maturity Model

Chapter 3 discussed the process of developing a strategic energy plan in detail. This section will focus on the content of the plan itself, the different areas of energy planning, and how to align priorities with industry best practices.

Specifically, this section presents the "energy program maturity model" as a holistic approach to energy planning. As it's used here, the term "maturity" relates to the degree of formality and optimization of processes, from ad hoc practices, to formally defined steps, managed result metrics, and active optimization of the energy management processes. The goal of this chapter is to help you think about each aspect of your energy plan in the context of industry best practices. This model should give you a basic benchmark against which to understand the maturity of your own energy program in each area.

There is no magic bullet to combat our energy challenges. The concept that we need to harness and leverage a wide variety of energy sources is true nationally, but also on an institutional level. It is necessary to consider the full spectrum of energy solutions when working to address the breadth of energy resilience imperatives that face your community. Identifying, analyzing, implementing and tracking all of these opportunities is a complex undertaking. Institutions need to move away from one-off energy projects led by whomever is available and think about a formal energy program, led by skilled, dedicated professionals.

This is not easily undertaken, but by breaking energy programs down into their key components, you can plan to progress in one area at a time. Results are certain to be mixed, but even uneven progress in multiple areas will result in an overall shift toward a more resilient energy system for your institution. This chapter describes the eight components of a successful energy program:

1. Goals
2. Strategy
3. Project Financing
4. People and Maintenance
5. Efficiency Technology
6. Energy Generation
7. Emergency Preparedness
8. Purchasing
9. Information Management

The model outlines characteristics found in five levels of program maturity—from no program through a best-in-class program, described below. For each area of energy planning, this section will provide example characteristics of what a program at that level might look like.

No Program

Institutions considered to have no program treat energy like a fixed cost and take no actions to address either the amount consumed or what the institution will do when energy supply is interrupted. Unfortunately, many institutions fall into this category. They are the hardest hit by supply interruptions or price increases, and their lack of active energy management may stifle their economic development.

Reactive Program

Institutions with reactive energy programs acknowledge that energy resilience has value and are willing to consider efforts on this front when they are presented. However, staff members do not seek out these opportunities, and there is no clear indication from management as to when such efforts will be funded. The organization is only prepared to *react* to energy opportunities as they arise.

Managed Program

This kind of program has guidelines around what kinds of energy efforts will be funded and even has goals around managing energy as a resource. These kinds of programs typically focus purely on cost savings and are not designed to evaluate projects on the broader criteria of security and risk avoidance. This is the kind of program found most often in corporations because it maximizes near-term economic outcomes.

Proactive Program

This kind of program recognizes that pursuit of energy resilience has long-term benefits that justify major investments. Proactive institutions work to identify energy resilience challenges and address them with planning and investment.

Best-in-Class Program

This category is reserved for high performers that execute the plan in a holistic, integrated manner. Achievement in this category may require considerable investment of time or money, but will result in programs that address all aspects of energy resilience, leveraging the strengths of the institution itself.

The purpose behind this model is to help your institution diagnose its maturity status in each of the areas below, so you can make an informed decision about where you would like to be and what it is going to take to get there. The priorities and internal resources of your institution will help you determine what level of maturity you would like to aim for in each area.

GOALS

As discussed in Chapter 3, setting clear goals and garnering buy-in from key stakeholders is a critical first step for a successful energy program. If there is no consensus around what an energy program is designed to do, it will be extremely difficult to get all players working toward the same ends. Goals accomplish three important things for your energy planning process. First, they provide simple language you can use to ensure that all parties involved are agreeing to the same thing. Simply put, clearly stated goals get stakeholders on the same page. Second, goals provide a sense of timing and scope around which plans and programs can be directed. Finally, goals provide metrics of success you can use to track your communities performance.

Goals should be created for each area of focus (such as generation, efficiency, and emergency response). Setting specific but realistic dates on these goals will help your team track progress and set priorities for a strategic plan.

The most critical aspect of goal setting is buy-in and accountability. Getting buy-in at the most senior level (mayor, CEO, base commander, etc.) will help to ensure the goals are taken seriously by the entire organization and are less likely to be superseded by a different effort. Having a senior leader endorse these goals will create the visible mandate necessary to motivate course corrections.

Of course, accountability is necessary as well. Achieving major goals can never be done solely from the top. Operators at all levels need to embrace actions associated with the energy program and make them a priority. The only way to ensure engagement is to identify individual roles in achieving these goals, and to hold these individuals accountable. A good way to accomplish accountability is to make those roles part of the annual review process. If achieving these goals (or some small part of them) is tied to the career advancement and compensation of individuals, the organization as a whole is likely to be motivated to realize its energy resilience goals.

STRATEGY

Managing energy usage and infrastructure is a complicated task. Changing the way people, equipment and funds are deployed for energy resilience is even more complex. The only way

Table 4-1. Goals

Maturity Level	Program Characteristics
No program	• No goals around energy
Reactive Program	• A general organizational focus on reducing energy waste where possible
Managed Program	• A goal to reduce energy consumption by a certain amount by a certain date
Proactive Program	• Defined goals to aggressively reduce energy consumption and improve resilience, often including a goal to increase utility management and on-site energy generation (e.g., renewables and others) by a certain amount and date
Best-In-Class Program	• Deadline-based, public goals that address energy use, energy generation, and the different operational aspects of the energy program • Energy goals are stated as part of the financial goals of the organization • As goals are met, they are updated

a community is likely to achieve even modest goals on this front is to develop a strategic plan. To put all this in context, let's use an example:

Institution A's stakeholders agree that energy resilience is important, and they agree on some goals around making their institution more energy resilient. They adjourn, and each manager does what they feel is most useful toward that end. While Institution A probably has some smart, dedicated professionals and their work is likely to lead to positive results, there is nothing to ensure that this team is working together toward the same specific ends; they are certainly not able to leverage each other's strengths and resources to create larger, collaborative results. What happens if various stakeholders have different definitions of what energy resilience is? What if stakeholders create competing proposals around different energy projects and both lose out because support for those ideas was diluted?

Institution B's stakeholders also agree that energy resilience is important and they agree on some goals around making their institution more energy resilient. Then, they develop a strategic plan that lays out a timeline of specific actions, who will be responsible for their completion, and what resources will be allocated to engender their success. Institution B can be sure that all players are working toward the same goals at the same time. This institution can leverage the strengths of various stakeholders to achieve impressive success in both the short and long term.

Developing a sound strategic plan with buy-in from all key stakeholders can be difficult, but is an excellent time investment for any institution looking to become more energy resilient. The steps outlined in Chapter 3 provide a basic approach for completing this kind of effort effectively.

PROJECT FINANCING

If a strategic plan is the brain of an energy program and superior technologies are the muscles, project financing is defi-

Table 4-2. Strategy

Maturity Level	Program Characteristics
No program	• No strategy for energy performance
Reactive Program	• No strategy for energy performance (reactive programs do not develop strategy documents)
Managed Program	• An energy reduction strategy is written at the Energy Manager/VP level and shared with facility management staff • The strategy focuses on short-term, quick-payback projects
Proactive Program	• A team of stakeholders draft a multi-year energy roadmap addressing all aspects of the energy program • Roles are well defined and lines of communication are open • There is strong support from upper management • Annual internal reporting against elements of the energy roadmap
Best-In-Class Program	• A team of stakeholders and senior management draft a multi-year energy roadmap addressing all aspects of the energy program • External partnerships are formed to further support strategy • Monthly internal reporting against elements of the energy roadmap • Annual or bi-annual review of the energy plan to update and recommit

nitely the blood. Having access to a steady flow of financing for cost-effective or high-priority projects is absolutely critical. Competition for internal capital can be so fierce that even energy projects with a one- or two-year payback can end up on the back burner. Even when project financing is available, there must be transparency around how to apply for or leverage this financing. Team members expected to identify and implement projects need to clearly understand the funding mechanisms and what information is needed by decision-makers to evaluate opportunities.

Energy programs that have a dedicated energy project financing mechanism tend to have the highest performance when it comes to identifying opportunities and capitalizing on potential benefits for energy resilience. The project financing market is expanding and becoming more sophisticated. Even institutions with little-to-no access to internal capital can now leverage several kinds of third-party financing. The trick to a healthy program is understanding the financing avenues your institution is likely to leverage and making those approaches part of your strategy from the outset.

One complicating factor for energy resilience efforts is the fact that some projects may have cost-savings benefits (like an efficiency retrofit), while others may not save money but will increase resilience (like system switches to keep certain buildings operating during a power loss). Ideally, an organization will have a budget to invest in energy resilience. If resilience-focused projects cannot be funded through separate capital dollars, the best approach is to bundle those projects with other, high-payback energy performance projects. This will result in a portfolio of projects under a single financing mechanism that, while achieving a slower payback, will result in greater energy resilience benefits. Managers may also advocate for special consideration for energy projects that have a resilience component. Perhaps these projects can be funded with slightly longer ROIs or can receive some other special treatment in the capital allocation process.

Table 4-3. Project Financing

Maturity Level	Program Characteristics
No program	• Any energy projects must be funded through each facility's capital budget
Reactive Program	• Energy projects can be submitted for internal capital and compete against other projects from other divisions • Emphasis on external incentives and financing
Managed Program	• There is a corporate internal rate of return (IRR) threshold set for energy projects • If they meet this IRR threshold, they are funded
Proactive Program	• There is a separate budget for energy investments • All energy projects are evaluated and the most attractive investments are paid for out of this fund • External financing is leveraged to fund some larger projects
Best-In-Class Program	• Energy savings are tracked and averted costs are put back into the energy fund to pay for future projects • Both internal and external financing are considered for all projects

PEOPLE AND MAINTENANCE

Too many energy programs focus on technology and completely ignore the human element. Effective management of energy-using equipment and preventative maintenance are equally important in pursuing a healthy energy management program and energy resilience. Without the guiding hand of an energy manager, it can be difficult for the various facilities management personnel to prepare for and respond to energy resilience challenges. Without an effective preventative maintenance program, vital equipment needed in emergency situations may be out of order or not performing at peak efficiency.

EFFICIENCY TECHNOLOGY

Deploying energy efficient technologies is typically a major thrust of any energy resilience plan. Reducing the total amount of energy used at an institution reduces risk from price spikes while lowering the total needs for on-site generation to combat outage risks. For this section, a program's placement on the maturity model is not about the use of specific technologies. Rather, it is about the systems your institution has in place for continually evaluating the technologies being used, planning for lifecycle costs as opposed to first costs, and understanding when retrofits will make sense in the context of your goals and requirements.

ENERGY GENERATION

The ability to generate energy on-site is critical to energy resilience. This means not only installing on-site generation, but being able to maintain it and, in some cases, fuel it. While some on-site generation sources (like wind and solar) require no fuel, other sources (like turbines or fuel cells) will require a steady supply of external fuel. When thinking about your program's maturity in

Table 4-4. People and Maintenance

Maturity Level	Program Characteristics
No program	• No one has responsibility for energy management • Equipment is fixed when it breaks
Reactive Program	• Facility management staff are expected to highlight energy saving opportunities they see, but energy consumption is not part of annual reviews • There is no communication between staff and management around energy • Preventative maintenance is directed at the discretion of each facility manager
Managed Program	• Someone within the organization is tasked with managing energy efficiency efforts (not full time) • Energy savings are considered as a positive performance factor for facility management in their annual review • Outside energy management courses made available to facility staff • There is a daily log book that facilities staff use to track preventative maintenance
Proactive Program	• A dedicated Energy Manger is hired • Goals around energy savings are set for energy management staff and possibly facility managers • There are clear lines of communication between energy staff and management • Meeting or not meeting these goals can affect salaries and bonuses • Energy management training is mandatory for facility staff • The organization considers energy efficiency experience in hiring decisions • There is a corporate preventative maintenance software used by all facilities
Best-In-Class Program	• A dedicated Energy Manger is hired at the VP level and reports to the CFO • Energy related goals are meaningful part of all energy stakeholder's (up to the VP level) individual performance plans and can affect salaries and bonuses • Various levels of internal energy-related training are available for both facility and non-facility-focused staff • Experience in energy efficiency is listed as a major evaluation criteria for new facility managers • There is an online tracking system for preventative maintenance and facilities staff are trained to use it

Table 4-5. Efficiency Technology

Maturity Level	Program Characteristics
No program	• No consideration is made for energy efficiency technologies • Lowest first-cost technologies purchased
Reactive Program	• Replacing end-of-life technologies with more efficient models • Technologies are installed without addressing systemic building issues
Managed Program	• Energy efficiency is incorporated into decision making process • Cost Benefit Analysis (CBA) is completed and technologies with under a 3 year payback are prioritized • Energy performance is considered when new buildings are designed
Proactive Program	• Energy audits of low-performing facilities are conducted in order to identify efficiency opportunities • High-value investments like variable speed drives and sensors are addressed • Load shifting technologies such as thermal storage are used to reduce demand costs • Energy savings is tracked for specific projects • Specific mandates for energy efficiency in new building construction
Best-In-Class Program	• Conduct energy audits of all facilities • Evaluate retrofits using lifecycle costing methods and net present value • Address technical retrofits that may have long-term payback but have other ancillary benefits • Use load shifting technologies such as thermal storage to reduce demand costs • Track energy savings for specific projects • Partner with other companies and national labs pilot cutting edge efficiency technologies • Develop building energy models for all new buildings to ensure that each design meets set standards for highly efficient buildings • Evaluate the water impacts of energy usage

the energy generation area, the first step is to consider how your institution is going about identifying and evaluating opportunities. This awareness of opportunities is a critical operation you will find in any institution making progress in this area.

Table 4-6. Energy Generation

Maturity Level	Program Characteristics
No program	• Limited to some emergency back-up power supply at critical locations
Reactive Program	• Organization considers on-site generation with a primary focus on up-time and power quality • Little-to-no attention is paid to renewables
Managed Program	• Organization conducts a financial analysis of opportunities to incorporate both fossil and renewable energy at new or low-performing sites • Incorporates on-site generation where feasible
Proactive Program	• Organization conducts a financial analysis of opportunities to incorporate both fossil and renewable energy at all sites • Develops pilot programs to test generation technologies/approaches • Some facilities are capable of becoming net-zero buildings
Best-In-Class Program	• Organization evaluates opportunities to incorporate both fossil and renewable energy at all sites – taking into consideration cost, risk/resiliency, and sustainability • Some facilities are able to generate more energy than they use and sell excess supply back to the grid • Collaborates with external organizations to conduct pilot programs to test innovative technologies and approaches

EMERGENCY PREPAREDNESS

Energy resilience planning is about preparing for the unexpected. The more an institution improves the energy efficiency of its facilities, the less backup power will be needed to keep them in operation. Mature generation programs will bring more energy resources to bear in times of emergency and allow some facilities to operate using on-site generation during emergencies that cause grid outages. There is a difference, however, between mitigating the effects of an undefined future energy shock through efficiency and generation projects, and engaging in an explicit emergency management process to identify specific energy risks and ensure that resources are properly aligned to contain and resolve them. More mature emergency management programs start with a detailed risk analysis and include rigorous emergency preparedness planning.

Table 4-7. Emergency Preparedness

Maturity Level	Program Characteristics
No program	• Organization does not include energy-specific information in its emergency management plans
Reactive Program	• "Power outage" is listed as one contingency in emergency plans • Checklist of staff actions in case of outage are included in emergency plans, including whom to contact to make the situation known
Managed Program	• A detailed risk analysis is conducted to evaluate potential threats and vulnerabilities to energy security on site • Policies and procedures are established for addressing an energy emergency and are mapped to identified risks
Proactive Program	• Emergency management is assigned as a collateral duty to one or more energy management staff members • Staff conducts an analysis of the financial impacts of identified energy risks • Elements of the energy system are prioritized for emergency response based on the institutional functions they support; the entire realized cost of a system failure in that area is estimated
Best-In-Class Program	• Energy management staff each have a well understood role in emergency situations • Energy management is an integral part of the organizations incident control system • Emergency mitigation projects to harden essential infrastructure and provide backup generation for essential functions are taken on based on risk analysis and financial modeling • All facilities and equipment are continuously commissioned with emergency preparedness in mind • All staff members are regularly trained on detailed emergency management plan, which includes energy assurance

PURCHASING

When attempting to reduce energy costs and exposure to price shocks, you have got to be a smart buyer. Institutions with no program or reactive programs tend to treat energy contracts as simply a cost of doing business. More mature programs treat energy purchasing as an opportunity to reduce costs and build in critical pricing protections. Your strategy may differ depending on your priorities—whether these are the reduction of typical operating costs or protection against price shocks. In the end, it pays to have energy purchasing professionals audit your current bills, understand your priorities, and aggressively negotiate with suppliers to yield the optimal purchasing arrangement.

Table 4-8. Purchasing

Maturity Level	Program Characteristics
No program	• Energy bills are treated as a fixed cost
Reactive Program	• When energy contracts expire, the purchasing manager looks for more competitive rates and considers longer-term contracts to lock in lower rates • When retrofits are made, the organization identifies applicable incentives or rebates
Managed Program	• Utility options are reviewed to determine best rates and schedule • Where possible, the organization combines the energy supply needs of different divisions in order to increase negotiating power • When retrofits are made, the organization identifies applicable incentives or rebates
Proactive Program	• Energy purchases are coordinated to increase negotiating power • Opportunities to engage in demand response and load shifting are identified • Purchasers are encouraged to purchase appliances with the ENERGY STAR® label • Pre-qualified ECM providers that can be quickly contracted to support ECM implementation are identified • The organization tracks energy-related incentives and rebates to determine when certain retrofits become financially optimal • Third-party billing analysis is conducted to ensure that energy consumption and pricing information is correct
Best-In-Class Program	• All energy purchases are coordinated through a central expert or team of experts • A combination of demand-response programs, load shifting and on-site generation gives facility managers flexibility to fully manage their energy sources and dramatically reduce demand charges • Standards in place to ensure all new appliances carry the ENERGY STAR label • Information tools are used to evaluate market options for ECM providers • Close coordination with state agencies and utilities to secure incentives and rebates that align with investment priorities • Third-party billing analysis is conducted on an ongoing basis to ensure that energy consumption and pricing information is correct

INFORMATION MANAGEMENT

While it is last on this list, information management should probably be your first priority when evaluating your program's maturity level. Measurement is the key to management and tracking how and where you are using energy in your portfolio is a critical prerequisite to developing an energy resilience plan. Institutions that find themselves in the no-program or reactive-program categories will be unable to set priorities in any of the other aspects covered in this section. Programs further along the maturity curve will be able to make smart purchasing and retrofit decisions, understand the weak points in their energy system, and respond much more quickly to emergencies or malfunctions in their energy grids.

As stated at the top, the goal of this section was not to suggest that every institution should strive to emulate the best-in-class behaviors in each section. Understanding your institution's place on the energy resilience maturity model is about taking stock of strengths and weaknesses in order to set priorities for future energy planning. Use the table below to summarize where you feel your own institution is along the maturity model in each area. This analysis can then be shared with your energy leadership team as part of the discussion around vision and priorities for your energy resilience efforts.

Energy Resilience Maturity Model

Table 4-9. Information Management

Maturity Level	Program Characteristics
No program	• Energy information is not tracked and is not readily accessible if asked for.
Reactive Program	• Monthly information on usage, demand and costs is collected centrally and available to managers who seek it out • Where required by law, buildings are benchmarked against a national database
Managed Program	• Daily energy consumption data is tracked by building • A BMS provides data on specific systems • Energy consumption data is benchmarked against national databases and updated on a quarterly basis • Performance data is available via a system all that facility managers are trained to use
Proactive Program	• Energy consumption is tracked by building in 15-minute intervals • A building management system and equipment controls allow remote management of individual system elements • Energy consumption data is benchmarked against national databases and automatically updated on a monthly basis • Portfolio-wide energy data is tracked against productivity to create direct energy-to-output metrics for the firm • Information tools are used to identify potential problems that could lead to energy waste • Key metrics are tracked for all levels of stakeholders in the energy chain • Performance data is included in regular reports seen by all facility managers and the Energy Manager • Energy numbers are used in GHG analysis in decision making
Best-In-Class Program	• Energy consumption is tracked in 15 minute intervals by submeters for each building's primary systems. • A BMS and equipment controls allow remote management of individual system elements • Monthly energy consumption data is benchmarked against national databases and automatically updated on a monthly basis • Energy data is correlated with each building's activities in order to track normalized energy productivity • Performance data is included in regular reports seen by senior management • Information tools are used to automate audits to identify potential ECMs • Key metrics are tracked for all levels of stakeholders in the energy chain and scorecards provide feedback • Public dashboards help building occupants to understand how their behavior affects energy usage • Energy consumption numbers are tied directly into automated GHG analysis

Table 4-10: Energy Resilience Planning Maturity Benchmark

Area	Current Maturity	Target Maturity	Current Strengths	Current Weaknesses	Potential Actions
Goals					
Strategy					
Project Financing					
People and Maintenance					
Efficiency Technology					
Energy Generation					
Emergency Preparedness					
Purchasing					
Information Management					

No Program
Reactive Program
Managed Program
Proactive Program
Best-In-class Program

Part III

Energy Resilience Performance Areas

Chapter 5

Energy Efficiency Approaches

Whatever method you use to develop your energy resilience plan, there will come a time to identify specific projects to pursue. Although energy efficiency and energy resilience are certainly not synonymous, there are three excellent reasons to look to efficiency and conservation first when developing any resilience strategy.

Reason One: Lower usage = lower risk

Using less energy means less risk exposure to various kinds of energy system shocks like short-term outages, short-term price spikes, changes in demand charges, long-term price increases and fuel supply interruptions. Reducing total energy-buy means energy consumption will have a smaller total footprint on your bottom line, and less potential to disrupt your organization's finances in times of higher prices. At the same time, stored energy (in both stationary and automotive fuel tanks) will last longer, allowing vehicles and backup generators to operate longer when fuel supplies are interrupted.

Reason Two: Decreased need for on-site generation

Reducing total energy load means less on-site generation and storage is needed to operate independently from the grid. As discussed in Chapter 6, energy resilience requires some sort of on-site generation. This infrastructure can be expensive. Reducing energy demand minimizes the size and expense of installing and maintaining on-site generation systems.

Reason Three: Efficiency savings can help fund investments

The cost savings associated with efficiency and conservation measures can help offset the cost of investing in new energy infrastructure like on-site generation, storage and transmission. Some organizations are willing to invest to improve their environmental sustainability. Many are willing to invest to reduce their risk of power loss. *All* organizations are willing to make investments in efficiency that will yield attractive paybacks. The parameters on those paybacks differ, but immediate financial reward from a good investment is always a motivator. As your organization weighs the value it puts on those first two priorities, you can counter real or perceived costs with the highly quantifiable savings from smart energy-saving measures. This kind of bundling can make a power storage or solar project revenue-neutral over a certain number of years when paired with efficiency investments.

The important message to remember here is to be expansive in your investment strategy, considering efficiency, conservation, generation and emergency preparedness as interdependent. Energy systems are just that—systems. Changes in one part can dramatically affect all other parts. As you develop your plan for energy resilience, it is possible to deliver value that is more than the sum of individual projects.

ENERGY EFFICIENCY AND CONSERVATION

For the reasons listed above, using less energy (or at least slowing the growth in energy consumption) must be considered the critical first step to any energy resilience effort. There are two key ways to manage energy use at your institution. The first is to promote energy efficiency—doing more work with less energy. Energy efficiency typically applies to the technologies we use: higher-MPG cars, more efficient lighting, buildings that retain heat and cold better, etc.

The other approach is energy conservation—finding ways to reduce the energy we need in the first place. This typically includes changing the way people operate. Conservation efforts are sometimes easier to engage in because they require less in the way of capital and technical retrofits. However, lasting results can be harder to attain in conservation because no central source can mandate them; they require changing people's mindsets and default behaviors. Still, there is much a community can do on this front, so it is important to consider both efficiency and conservation strategies when working to reduce energy demand.

This book is not going to attempt to describe each demand reduction technology or best practice available today. Any attempt to do so would be incomplete and almost immediately out of date, as improved technologies and solutions are constantly being developed. Instead, we will focus on best practices for identifying, evaluating and comparing demand-reduction approaches.

TAILORING DEMAND REDUCTION TO MEET RESILIENCE GOALS

The technologies and strategies you use to improve energy efficiency should take all of your goals into account. Cost savings and environmental performance are often top priorities of energy efficiency projects. As such, projects that have the largest impact on total energy usage and the quickest payback are often taken on first. However, in the context of energy resilience, the "low hanging fruit" approach may not be the wisest course of action when choosing which projects to implement first.

In a crisis (whether an outage event or other problem that limits access to energy), certain technologies will be critical. Lighting, heat and, in some cases, refrigeration will be top priorities. Mechanical equipment, such as pumps or communication equipment like telecommunications switching stations, may also be vital. As part of your overall resilience plan, prioritize buildings and specific end-uses for energy so that you can develop a triage system to meet the most critical loads first. This concept is discussed in more detail in Chapter 7. The prioritization of end uses at your institution can inform your decisions around where to perform energy audits, and which end-uses to target with efficiency and conservation measures.

As an example, let's look at two retrofit opportunities for a community that has some on-site generation, but not enough to run all of its facilities in the case of a grid outage. A public library needs upgrades to its boiler and chiller. Both are working, but are old and inefficient. The project would have a total cost of $45,000 with a payback of 3.5 years. Meanwhile, the county could also spend that $45,000 upgrading the lighting in the police station, fire station, community center and City Hall. These buildings already have relatively efficient T-8 lighting, but an upgrade to LED lights would save a great deal of energy, and have a payback of five years. If the goal of energy management is purely about cost savings, the boiler and chiller upgrades are the better project. However, heating and cooling the library will be a lower priority during a grid outage than lighting government buildings essential to emergency response. Given this community's limited on-site power options, lowering the total energy use of that lighting is a higher priority than the energy usage at the library. They are both good projects with a solid payback and, hopefully, both can be implemented. If one has to be chosen, though, the lighting upgrade will do more to improve energy resilience.

CONTINUOUS ENERGY PERFORMANCE IMPROVEMENT

Goals around energy resilience are not required to justify energy efficiency as a management best practice. The ideal approach is a portfolio-wide analysis to identify improvement opportunities, implement top priorities based on available budget, and track performance. Any organization that follows the nine basic steps listed below will achieve what we call *continuous energy performance improvement*. As equipment fails, technology improves, or needs change, your facilities will continue to perform better because, like a high-performance car, they are continually tuned-up.

You will notice some overlap between this approach and the higher-level approach discussed in Chapter 3. Energy resilience planning is a broader effort than simple energy management, but several of the best practices are similar and the work done in the energy management area would logically roll up to your larger energy resilience planning efforts.

Nine Steps to Continuous Energy Performance Improvement
1. Portfolio-wide benchmarking
2. Targeted building energy audits
3. Targeted building commissioning and re-commissioning
4. Identification and prioritization of energy conservation measures (ECMs)
5. Determination of financing approach
6. Selection of energy efficient technologies to be deployed
7. Implementation of ECMs
8. Performance tracking of ECMs
9. Repeat Cycle for Continuous Improvement

The section below discusses each of these steps in more depth.

Portfolio-wide Benchmarking

The very first step in an energy management program should be the performance benchmarking of all facilities in your portfolio. This energy performance benchmarking is a large portion of what was discussed in Chapter 3, Step One—though in this case we are not concerned with generation sources and outage preparedness. Here, we are concerned only about using as little energy as possible in our systems.

Developing an energy baseline allows you to understand the starting point against which you will set goals and measure progress. By taking historical utility data and analyzing the implications on a building's energy usage, you can make some important assumptions about your opportunities

Critical elements of energy performance benchmarking include:

- *Inclusion of all energy sources* to ensure that the full energy picture is evaluated. It is also highly recommended that water utilities and greenhouse gas footprint are included here for a complete view of utility impact.

- *Weather normalization* to create apples-to-apples comparisons as we compare building performance across years or across buildings.

- *Analysis of results* to identify which facilities have abnormally high energy use intensity per square foot (EUI) for their type; this analysis also indicates which buildings are unexpectedly increasing in EUI (often due to changes in operations or equipment malfunction).

The most common approach to energy performance benchmarking is to enter three years of historical building utility data into performance benchmarking software. ENERGY STAR's Portfolio Manager (www.energystar.gov/benchmark) is an excellent and free online tool for energy performance benchmarking. It will provide figures for weather-normalized energy intensity per square foot for each building in your portfolio, and will also give you a 1-100 score for most space types, letting you know how efficient that facility is compared to others like it across the country (again, weather normalized). There are other software products for sale that offer the same benchmarking capabilities, but combine them with more robust analytical functionality.

Whether you use Portfolio Manager or another tool, the important thing is that you look at energy intensity per square foot and that your analysis is normalized for local weather. Simply tracking energy costs per building in a spreadsheet can be misleading. One particularly cold winter or hot summer could mask major improvements in energy performance.

A solid performance benchmarking effort is essential to understanding how and where your organization uses energy. It allows you not only to identify poor performers that will be good candidates for in-depth energy audits, but also to identify highly efficient facilities that may be employing best practices that should be shared with the rest of your portfolio.

Figure 5-1 shows a standard curve distribution for energy performance based on Portfolio Manager's 1-100 rating scale. As you will note, the majority of your buildings are likely to have av-

Energy Efficiency Approaches 111

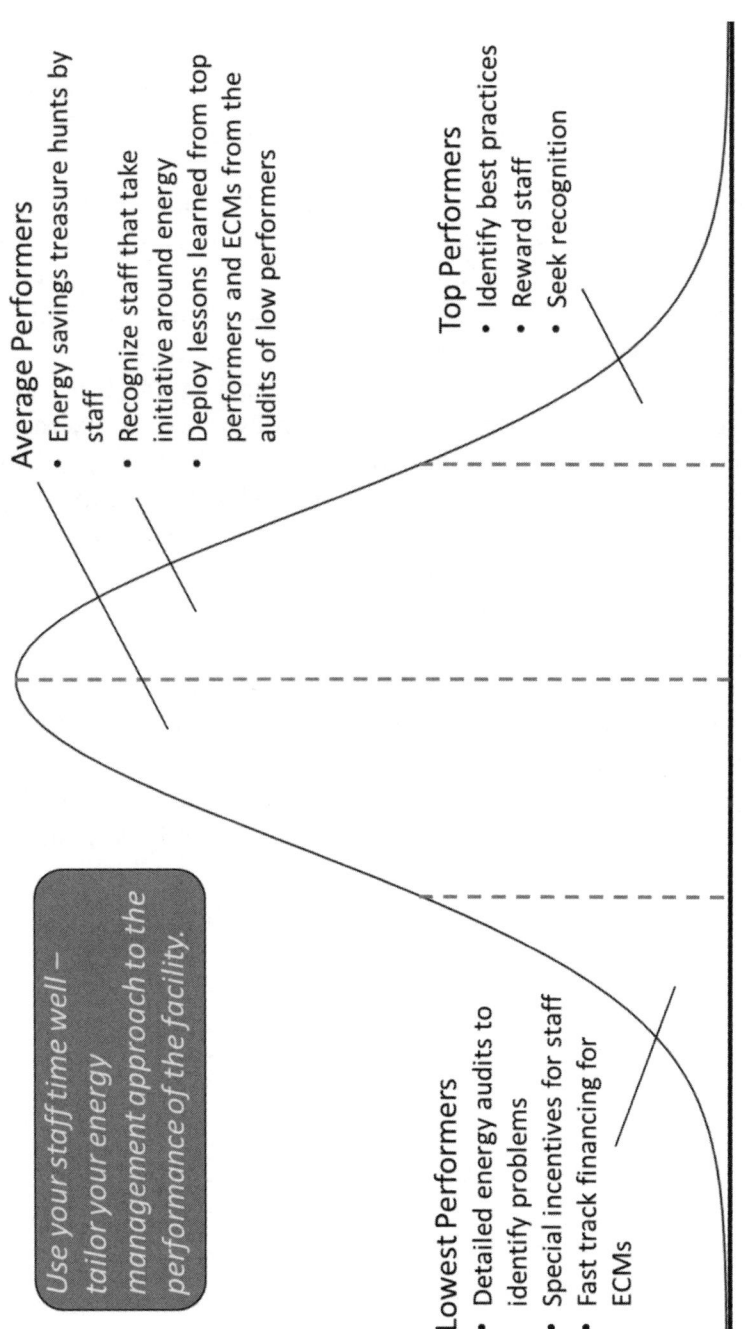

Figure 5-1: Energy Performance Benchmarking Distribution.

erage performance while a small subset will be in the bottom and top quartiles. It is at these facilities where you can glean the most lessons learned for your portfolio. The low performing facilities should receive in-depth energy audits to identify operational and technical problems that may be systemic throughout your portfolio. This will give you example projects, case studies, and practices that you can share across all of your facilities. Interviewing the facility managers for top performing buildings may provide insight into construction, equipment, or operational best practices you should deploy across the portfolio.

Advanced energy managers benchmark more than just energy usage. They benchmark all aspects of their energy management situation. Studying the current state of operating procedures, the skill sets of your team members, technology purchasing guidelines and even maintenance procedures will give you a full understanding of the mechanisms driving your organization's energy performance. This knowledge is particularly useful when combined with energy performance benchmarking, and it will give you the data points you need to understand the correlation between certain behaviors/technologies and energy performance.

Knowing your energy baseline goes beyond simple benchmarking, especially in the context of energy resilience. The table below outlines other aspects of your energy system to consider in your baseline. Your goal is to understand where your opportunities lie so you can scope out solutions and evaluate those as part of your plan development process.

Table 5-1 provides some different energy focus areas and some questions you should be asking about current operations and potential solutions. (Note that this is not a comprehensive list, and that you should attempt to flesh this out with additional questions germane to your institution.)

Figure 5-2 shows one way of thinking about portfolio-wide energy performance improvement. It presents a few key points discussed in this chapter:

- Analysis is a critical first step before any planning can take place.

Energy Efficiency Approaches

Table 5-1

Focus Area	Key Questions	Example Actions
Energy Purchasing	• What rates are you paying for energy? • What is your risk exposure if prices jump? • Are you only paying for the energy you use?	• Compare bills against your known portfolio to ensure you are not paying for utilities you are not using • Evaluate your power purchase agreements • Consolidate your energy purchasing negotiating power in order to get the best rates • Evaluate options to engage with different energy suppliers (if you live in a deregulated state)
Demand Charges	• How much do these affect your energy costs? • Based on your projected operations, are these expected to increase?	• Identify load-shifting techniques you can employ to flatten your demand profile • Identify utility programs and incentives around demand response
On-site Generation	• What kind of back-up power or on-site generation do you currently have? • How long will this last if you lose grid connectivity? • What loads can this generation equipment handle?	• Identify sites where additional generation could potentially be installed • Determine which renewable energy sources are potentially available [The National Renewable Energy Laboratory has resource maps you can explore to better understand your local potential: http://www.nrel.gov/renewable_resources/.]
Technology Efficiency	• Do your facilities employ any considerably out-of-date technologies? • What does your benchmarking tell you about relative building performance?	• Develop a facility auditing program to identify technologies that require retrofit • Engage energy service companies to identify and deploy retrofits
Operating Procedures	• What are the start and stop times for HVAC in most facilities? • What are the temperature set points for heating and cooling? • Are lights being used only when needed?	• Evaluate best practices from across your portfolio • Share these best practices across facility management staff • Develop a common set of facility management guidelines
Team Skills	• Of the staff controlling your facilities, how many are PE's, CEMs, etc.? • Which facility managers have taken specialized training in energy management?	• Develop guidelines for energy management training for facility staff • Engage non-facility staff with training on how they can help conserve energy
Project Financing	• How have energy performance measures been funded in the past? • What is the process for facility management staff to request funding for identified projects?	• Create clear understanding around the process for proposing projects and obtaining funding • Have a clear policy describing the use of performance contracting • Create a revolving energy project loan program

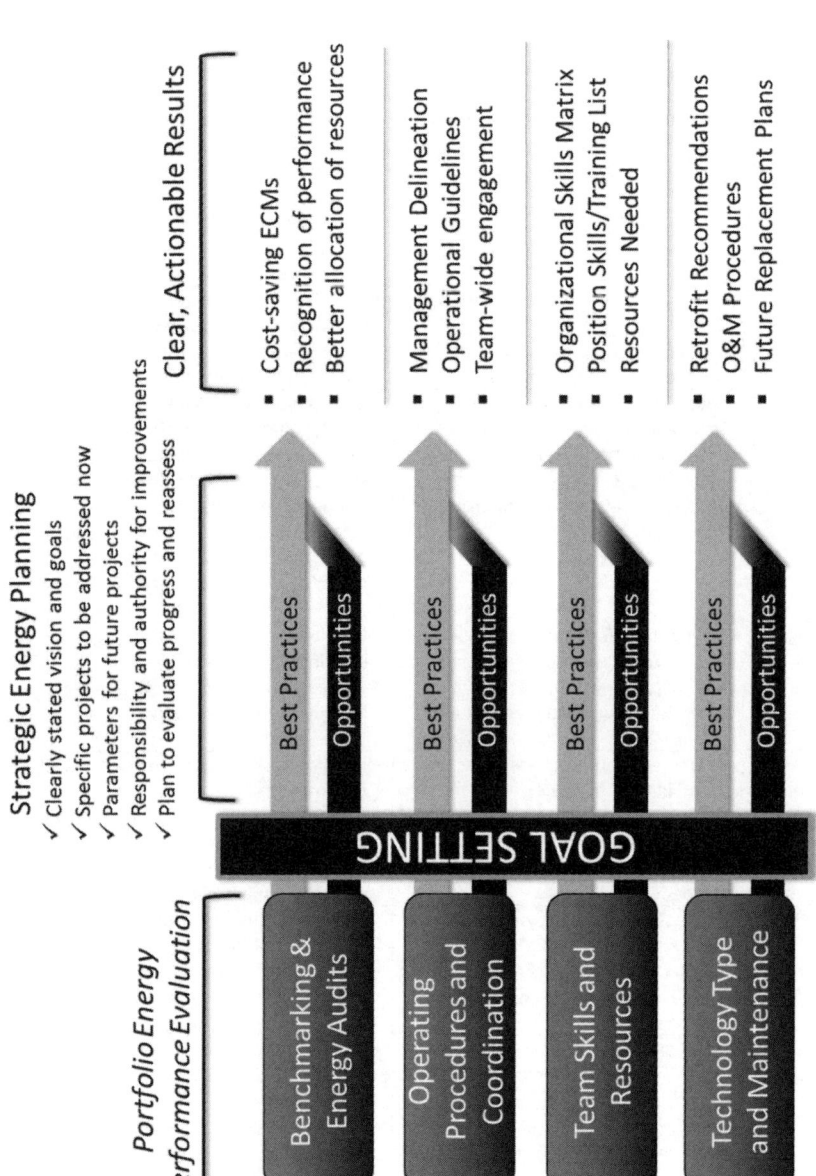

Figure 5-2: Energy Performance Evaluation of a Building Portfolio

- Once improvements have been made, a proactive performance management effort is the only way to ensure that performance continues.

- Opportunities can be found beyond just technical upgrades; you should also explore the opportunities inherent in new operating procedures and team skills/resources.

Targeted Building Energy Audits

The purpose of an energy audit is to identify opportunities for energy savings based on the current energy profile at your institution. These opportunities are referred to as energy conservation measures (ECMs). The Energy Independence and Security Act of 2007 requires Federal agencies to designate covered facilities that account for at least 75 percent of the agency's energy usage and then to perform energy audits on those facilities every four years. For larger private portfolios, this Act is probably a good approach to emulate—as those facilities with the highest energy use should also be the highest priority for audits. For facilities where a full on-site audit won't be cost effective, there are alternative options to do remote energy performance opportunity analysis through interviews with staff or even using energy data analysis software.

Not all energy audits are created equal, and you will want to ensure that any audit you undertake meets your specific needs and desired level of detail. Building effective audits usually requires recruiting a qualified professional to evaluate the energy use of all equipment and related operational procedures. Hiring an outside analyst can expedite your research and ensure that you are choosing the right variables to explore. Your local utility may be able to recommend reputable firms that provide this service. Some utilities even provide financial assistance to help customers conduct energy audits.

Before commissioning any audit, you will want to get familiar with the ASHRAE (formerly the American Society of Heating, Refrigeration and Air Conditioning Engineers), tiered classification system for energy audits.[71] These categories include:

ASHRAE Level I Audit: This audit—frequently called a "walk through" or "screening" audit involves just that—walking through the space to identify areas of energy waste or inefficient technologies. This kind of audit does not involve actual measurements and does not result in any financial analysis of the ECMs it identifies. This audit is best used to gather basic information about what opportunities exist in your portfolio. That information can then help you direct funding for more in-depth analysis like a Level II or III audit. This approach is not very valuable for making ECM investment decisions, however, because of the lack of detail and financial analysis involved.

ASHRAE Level II Audit: A level II audit is a more involved process that evaluates energy performance opportunities in the building envelope, lighting, HVAC, domestic hot water, plug loads, compressed air and process uses (for manufacturing, service, or processing facilities). The audit team collects data about the technologies, controls, and management practices in place. They then use this data to perform financial analysis around each ECM. A level II audit will provide you with ROI and simple payback for each ECM identified. This level of audit is typically sufficient for organizations looking to identify and prioritize ECMs to implement with internal funding.

ASHRAE Level III Audit: Level II audits provide an estimate of energy savings, but do so using common assumptions about energy performance. When significant amounts of capital are required for a retrofit, a greater level confidence is needed before making that investment. Level III audits, or "investment grade audits," typically look at specific ECMs identified in a lower-level audit and conduct much more in-depth analysis of the anticipated effect of that ECM. This analysis is done using computer models of the building to accurately simulate the impacts of new technologies or practices. Creating these computer models often includes the placement and tracking of data loggers and calibrating the model to ensure it operates the way the building does. These audits are expensive and only recommended for institutions considering large investments.

Be sure to think carefully about which kind of audit will meet your needs. Spending money on a series of Level I audits will only be a good investment if a survey audit will provide enough information for you to move forward.

In addition to (or sometimes instead of) a formal audit from a third party, some firms use internal energy treasure hunts to identify energy saving measures. These are essentially Level I audits performed by your internal staff and only require setting aside time for facilities and maintenance personnel to evaluate energy using equipment and procedures and write up recommendations for potential fixes. The advantage to this approach is that your team knows their buildings well, and have seen them operate over time. The disadvantage is that your team members are probably not trained to identify and evaluate ECMs. They have also seen things work for a certain way for so long that they may not consider alternate approaches. Treasure hunts are a great way to tap great ideas from your staff—particularly between audits—but they are not a replacement for an evaluation by a qualified auditor.

Finally, it is important to remember that performing energy audits does not reduce energy consumption. It simply provides insight into where energy waste can be found at each facility. If the report finds its way directly into a desk drawer, the audit was a waste of money. If the organization is not prepared to identify funding to implement cost-effective or resilience-improving concepts, then the audit is wasted effort. You should be prepared to have high-level meetings to discuss the recommendations that come out of each energy audit, and determine who will lead implementation of the ECMs selected. You should also set up a follow-up meeting, three to six months down the line, to track progress and maintain accountability.

Targeted Building Commissioning and Retrocommissioning
For some buildings, one of the most cost-effective ways to

improve energy performance is through retrocommissioning. When a building is first brought online, smart facility managers put it through a commissioning process. This is an analysis to ensure that all energy-using systems in the building work properly. Even LEED Platinum buildings can experience problems in construction or controls design. A commissioning effort identifies where systems may not be operating as they should. By extension, retrocommissioning is simply performing a commissioning study on an existing building. This kind of study can catch a number of energy-draining problems like incorrectly installed equipment, poorly calibrated controls, incorrect overrides, and failing equipment. Unlike an energy audit, which focuses on sub-optimal processes and technology, a retrocommissioning study includes more in-depth testing designed to identify where the building is not operating as designed.

Whether you identify the need for retrocommissioning from an energy audit or simply decide to make this part of your regularly scheduled efforts to improve energy performance, these efforts have been proven to yield results. A study by the Lawrence Berkley National Laboratory found that "re/retro-commissioning yields a median 16 percent energy savings with a payback time of 1.1 years for a cash-on-hand return of 91 percent."[72]

Policy makers have recognized the power of retrocommissioning and begun writing it into laws that affect both government and private portfolios. Policies like the Energy Independence and Security Act and Executive Order 13514 set clear standards for new energy management efforts at Federal facilities, and specifically name retrocommissioning as a tool to be used to improve energy performance. [An excellent summary of Federal legislation around retrocommissioning can be found in "Re/Retro-commissioning: The Best Kept Secret You Can't Afford Not to Know," Thompson, Woody and Moore, Ben, Strategic Planning for Energy and the Environment Vol. 31, No. 2, 2011] New York City has, with Local Law 87, required all buildings over 50,000 square feet to complete an energy audit and retrocommissioning study. While this is currently the most aggressive local law

ENERGY EFFICIENCY APPROACHES 119

around energy management in the United States, New York's Local Law 84 requiring building benchmarking was eventually copied and implemented by several other cities and the state of California. It is likely that other local or state governments will soon begin adopting something very similar to Local Law 87. [A useful summary of these laws was developed by The Institute for Market Transformation and is titled "State & Local Energy Performance Regulations." It can be found at http://www.imt.org/] It is not a stretch to expect a similar proliferation of laws requiring retrocommissioning.

Identification and Prioritization of
Energy Conservation Measures (ECMs)

Once you have completed your analysis, you will have a list of measures to consider. If you performed a Level II or higher audit, you will also have financial information around what those projects will cost and their probable payback. It is likely that you will need to select some subset of these ideas, due to constraints in budget or staff time to manage projects. This step is, of course, the same as Step Six in Chapter 3 and you may be also be weighing your energy management projects against projects in the on-site generation or emergency planning areas.

Your first priority is to set the parameters around which you will evaluate and rank projects. Is there a specific return on investment (ROI) or payback threshold that projects need to meet in order to obtain funding? Are there buildings slated for renovation or demolition in the foreseeable future, making long-term investments a non-starter? Are there projects that, if bundled together, may yield additional savings? One example may be coupling a building envelope insulation project with a boiler replacement—if the tighter building can be well served by a smaller boiler system. As discussed above, simple payback or ROI is a common top driver, but using this as your sole criterion may not result in an optimal outcome for energy resilience. Consider the full range of drivers and limitations before selecting a set of ECMs for implementation. It is also important to note that most energy performance calculations will involve a mixture of data and estimates. The more information that is estimated, the higher the degree of uncertainty around the figures produced. Include some awareness of how firm each projects numbers are when comparing proposals.

In his article, "Where Energy Efficiency and Alternative Energy Work, Where They Don't, and Why," Donald Wulfinhoff outlines ten questions an organization should ask itself when considering new energy efficiency projects or programs.[73] Note that he also introduces a new parameter he calls the "energy return ratio." This is the ratio of the energy saved/generated by a project to the energy required to build and implement the project. Wulfinhoff's ten questions are:

1. What is the energy return ratio of the project?
2. How does it work?
3. How close to reality is it?
4. How much savings or new energy can it provide?
5. Can it work well in our location?
6. What is the most efficient scale?
7. What other benefits does it offer?
8. What problems does it have?
9. Do we have the management attention and staff skill to operate it successfully?
10. Where else could we invest our resources more profitably?

Answering these questions for each of your potential projects will ensure that you have thoroughly evaluated their strengths and weaknesses. It is then up to you to apply that knowledge to your organization's priorities around energy costs, resilience, sustainability, and public image.

Determination of Financing Approach
This step should take place simultaneously with the prioritization of ECMs since the funding mechanisms used will affect

what is possible to pursue. Refer back to Step Seven in Chapter 3, which discusses some of the main funding sources available today. One common challenge in this work is that the energy professionals that understand the ECMs are typically not very fluent in the finance side of the equation. This leads them to mistakenly believe their maintenance operating budget is the beginning and end of their financing opportunities. Conversely, the financial professionals within many organizations do not understand the energy side of the equation, and frequently overestimate the uncertainty involved in performance retrofit projects. The reality is that there are several potential funding streams for ECM projects, and at least some education is likely to be needed to bring all stakeholders up to speed on what needs to be done.

There are other funding sources unique to specific organization types (like institutional endowments or bond issuance) not discussed in this book. Whatever funding approach your organization takes, be prepared to have concrete estimates of anticipated energy savings (with clearly stated assumptions) before determining your financial approach. You may need to make your case to a variety of audiences. Energy champions often report that getting projects funded is the most difficult part of their jobs. It is important to remember, however, that initial success will likely generate more success. If you can demonstrate that ECMs implemented in the past have lived up to predictions, funding new projects will be that much easier.

Selection of Energy Efficient Technologies to be Deployed

Once you have determined *which* ECMs you will implement and *how* you will pay for them, it's time to determine which specific technologies or manufacturers will be selected for purchase. There is no magic bullet here or single source for all the information you will need. Your first step is going to involve tapping the subject matter experts you have access to—whether they are internal staff, engineering partners or outside consultants. These people typically read about and become familiar with new technologies as they become commercialized. In addition to these resources, there are excellent studies conducted by prominent labs around new technologies. Labs such as Lawrence Berkeley National Laboratory, Pacific Northwest National Laboratory, and the National Renewable Energy Laboratory conduct outstanding analysis of new technologies with no bias toward or against specific product types or manufacturers.

Consider carefully before committing yourself to a specific technology. If you have inefficient T-12 fluorescent lighting, does it make sense to upgrade to T-8, T-5, or LED? If you choose LED, should you do a complete lighting retrofit or simply select LED lights that can plug into your existing fixtures? Should you also consider some kind of smart lighting sensor system to pair with this retrofit? And do the answers to these questions change when you consider not only payback but also energy resilience? You will need some detailed analytics to determine if you are spending retrofit dollars on the right products. If you don't have in-house capabilities to provide this kind of analysis, create your own team by engaging experts, e.g., partners such as a consultant, an engineering firm, or even your utility companies. When planning large-budget retrofits, strongly consider an ASHRAE Level III audit focusing on those technologies.

Finally, as with all major purchases, it is important to engage multiple potential vendors—regardless of whether you plan to make your purchase through a proposal process or by buying directly from a single source. Understand how their products are installed, what kinds of warranties they offer, how their performance may differ from what you currently have, and what additional training or maintenance work will be required of your facilities staff to operate them.

Implementation of ECMs

Implementation of ECMs is a critical project management step, and can be more of a challenge than expected. Various contractors, manufacturers and facilities team members must be coordinated to ensure the right people are installing the right technologies at the right facilities at the right time. There is no magic formula to making this work. Here are a few recommendations based on our experience with these projects:

Put a professional project manager at the helm—Too many engineering students graduate without hard skills in project management. Understanding the technology being implemented is not enough to qualify a person to lead a major retrofit project. Select a project manager with real experience—someone who is familiar with Gantt charts and resource management. Consider assigning a more junior staffer to participate as a learning opportunity to ensure that you have competent project managers available as you develop future projects.

Incorporate third party installation into project costs—It's not uncommon for facility management and maintenance personnel to report that they spend 120 percent of their time putting out fires. Organizations that buy new energy efficient technologies with the expectation that staff members will implement them in their spare time find themselves with closets and warehouses full of very efficient, and not-at-all useful technology. It makes much more sense to simply incorporate the cost of hiring experts to install new technologies. A slightly more expensive ECM that begins saving you money right away is vastly preferable to a slightly less expensive project that never actually gets installed.

Create specific deadlines and incorporate them into your contracts—I there is no deadline, there is no incentive to complete the project. Your suppliers and installers need very specific guidelines built into the contract with penalties for any delays they cause. Every minute your new project is not online is will cost you money. If your vendors are not going to put you at the top of their priority lists, they should share in your financial losses.

Incorporate completion of these projects into the performance evaluations of key staff—Project delays are not always the fault of third parties. Your own staff needs to be invested in getting ECMs implemented completely and on-time. Create career incentives for implementing these projects well and disincentives for poor or delayed work. This will lead to better engagement within the team, and it will give your organization an excellent opportunity to gauge project management and leadership skills.

Performance Tracking of ECMs

Once implemented, you should closely track the performance of new energy technologies or procedures. Tracking is critical for two reasons. First, adding new technologies or approaches into your system can have unanticipated consequences on other systems if done incorrectly. Tracking the performance of your new systems is a way to ensure that you get what you have paid for. Second, demonstrating the cost savings from ECMs is an excellent way to build internal support, which can lead to further investment in energy management and demonstrate the value of your team to cost control.

Tracking ECM performance has not always easy and straight forward. At the very least you should continue to benchmark energy performance of your facilities after a retrofit. Absent other factors like a change in building usage, you should expect to see a drop in weather-normalized energy use intensity after implementing ECMs. If you have completed a major ECM project, consider installing data loggers to track energy used by specific equipment. This level of analysis requires some additional investment, but gives you the highest level of resolution possible to evaluate performance of your new systems.

Repeat Cycle for Continuous Improvement

When an organization is lucky enough to have a team of energy champions to implement this kind of analysis, prioritization, implementation and tracking effort around energy performance, the most common response is, "Whew, we're done!" However, a successful energy management program must be based on the goal of continuous energy performance improvement. Buildings age and become less efficient. The technologies available in the

Energy Efficiency Approaches 125

marketplace improve and create new opportunities for savings. Best practices may be lost in personnel turnover and need to be readdressed.

Once the ECMs identified in your analysis phase have been addressed or consciously set aside, it is time to begin again. This should not mean performing energy audits every year. Indeed, the cycle to fund and implement new ideas is typically longer than one year. It does mean that your team should constantly be spending some of its time on one of the eight steps above for each facility. Some of them, like performance benchmarking, should be ongoing efforts.

The good news is that there are ways to lower the burden on your team and improve your internal energy management skills over time. Institutions managing larger portfolios should consider staging facilities so only a portion of them are being audited this year. If you have a portfolio of 100 buildings, auditing 20-25 each year will result in a manageable process. A lower-effort continuous process is much more likely to create sustainable success than a one-time major initiative that requires staff members to take time away from their other typical duties.

If you are able to truly incorporate these continuous energy performance improvement strategies into your team's standard operating procedure, it will get easier. You will develop the internal resources and skills to make these projects happen more quickly and efficiently. By mentoring more junior staff on the evaluation, analysis, and project management elements, you will increase your energy management capabilities while providing for professional growth within the team.

PRACTICAL SOLUTIONS FOR ENERGY PLANNING

A successful plan needs to recognize a basic hurdle: Most facilities management professionals already have very full plates. To ask your team to add new energy management activities to their responsibilities without additional resources is to invite failure. Unless your team is already practicing the steps of continuous energy performance improvement, implementing a program like this is going to require time and attention—bandwidth that simply may not be available right now. Your organization, at a high level, needs to recognize the value of energy performance improvement for both the bottom line and the energy resilience of the community. As part of that recommendation, consider the following measures:

Hire or Designate an Energy Manager

This person's primary role should be to manage energy generation resources, reduce energy usage, negotiate energy purchasing contracts and generally lower energy costs. This person shouldn't also be the person people call when a boiler fails. They should not also be the environmental, health and safety coordinator. If you have a smaller portfolio and a smaller budget, you are better off reducing the cost of this role by hiring a more junior person for it. At the end of the day, these efforts require a full-time champion who can evaluate the data and bring good ideas to the attention of senior management.

Create Visibility with Senior Management

The best energy management programs are integrate upper management in oversight and performance tracking. Energy resilience projects can be very difficult to achieve without high-level support and buy-in. Scheduling monthly or at least quarterly meetings in which the Energy Manager updates senior staff will ensure that all parties are held accountable for achieving project goals, and present a forum for discussing next steps.

Carve out Time for Other Team Members to Support the Energy Manager

A champion is critical, but a team of champions is far superior. While most organizations can only dedicate one position to energy management, it makes sense to create opportunities and incentives for other team members to participate in energy resilience efforts. World-class energy management is not something one person does—it is part of the culture of the organization. Also, do not limit this effort to the facilities team. The most effective energy programs leverage skill sets and knowledge from numerous departments. Finance, operations, marketing—pretty much every team can contribute to this effort.

Reach out to Find Partners

Most leaders reading this book will have stakeholders outside of their own organizations. Energy resilience efforts do not need to be planned and executed in a vacuum. Whether it is technical support from the local utility, in-kind support from a business group, or project cost-sharing with a neighbor that can benefit from the same project—engaging partners is an outstanding way to add resources to the project and amplify benefits. The Fort Carson case study (below) is a prime example of engaged partnership.

Publicize Your Successes

Whether you're only reducing energy usage or you're deploying a broad energy resilience effort, your work has an impact beyond your own finances. It impacts the environment, the community, and the marketplace. Even if your organization is not doing anything that is truly cutting edge, you are making positive changes other people will admire. Use internal communications, annual reports, client newsletters, social media and other channels to communicate new initiatives and proven successes from past initiatives. Any positive attention from this communication will create goodwill for the organization and build internal support for further energy resilience efforts.

Continuous energy performance improvement cannot be a one-time activity. It can't be only one person's job. While it can be supported by a variety of professionals, your organization cannot outsource this function entirely. Energy management has to be integrated with your organization's strategic and business planning efforts. It needs to be integrated with the resilience efforts of on-site generation and emergency planning. To some extent, it needs to be part of your organizational culture and involve every member of your organization.

> **Case Study—Fort Carson[74]**
>
> Fort Carson is a U.S. Army base in Colorado. They do not have to worry much about earthquakes, hurricanes or tornadoes. Their electricity is run through three separate sub-stations providing excellent redundancy when it comes to grid reliability. Why, then, is Fort Carson a clear leader in energy resilience planning? In 2002, the base partnered with the local community to develop a 25-year sustainability plan with 12 distinct goals. One of those goals was energy-use reduction.
>
> Hal Alguire, the Director of Public Works at Fort Carson believes they can reduce energy consumption 10-30 percent based solely on engaging the staff and soldiers. Recognizing that the people on base drive energy usage, the energy management team started by looking at smart meters. By having better, more granular data on energy use, they were able to share that data with commanders and establish energy saving goals for individual brigades.
>
> In conjunction with engaging personnel, the base has aggressively pursued energy efficiency savings through technical retrofits. They have completed several projects in lighting, high-efficiency boilers and HVAC. The base began using LEED EB as a standard for retrofitting facilities, and they now have over 50 LEED certified buildings. All this certification was done in conjunction with the local Army Corps of Engineers office, and the Army now makes LEED certification a requirement for all its new buildings nationwide.
>
> While pursuing efficiency, the base is also piloting new on-site generation approaches for the Department of Defense (DoD). Ft Carson is part of a micro-grid demonstration project called SPIDERS (Smart Power Infrastructure Demonstration for Energy Reliability and Security). They have gas generators and five electric vehicles as part of the demonstration program. The SPIDERS

Energy Efficiency Approaches 129

> program is in partnership with the National Renewable Energy Laboratory, Sandia National Laboratory, an automotive company and U.S. Northern Command. Fort Carson also has a 2-megawatt solar array, 1.5 MW of distributed solar panels and is currently conducting studies on the potential to deploy wind power.
>
> Since developing this plan, the base has reduced energy intensity (kBtu per sq ft) by 17 percent. The money saved on utility bills is being used to help fund other efforts in the 12-goal plan. The Fort Carson approach demonstrates two key characteristics of a successful program. First, the team is not simply looking at technology or behavior—they are addressing both and letting the efforts in each area inform the other. Second, in each effort they have pursued, the Fort Carson energy team has leveraged entities outside its facilities team—the soldiers on base, corporate partners, the local community, NREL, and other programs at DoD—to create momentum and highlight the strides the base is making in reducing energy consumption and increasing energy resilience.

Fort Carson's success was a combination of energy efficiency measures and on-site generation. As you start to understand your post-efficiency-effort load profile, on-site generation will be the next logical step. The following chapter offers insight into the on-site generation planning process, major technologies to consider, and the value of developing an integrated microgrid.

Chapter 6

Community Energy Generation

When the power stops flowing, it doesn't matter how efficient your facilities are, or how careful your emergency procedures, your community will still be at risk. That risk increases the longer you are without power. Having on-site power and the ability to deliver it where it is needed is the critical foundation of any energy resilience plan. This chapter will discuss how you evaluate your power needs, review technologies communities are turning to today for on-site generation and power storage, and describe the impressive potential of on-site microgrids to maximize the value of those technologies.

Before the development of the modern American power grid, electric power in the United States was generated near the point of use—first with rudimentary wind and hydro power and later with small coal-fired engines. As industrialization and the electrification of homes became driving forces of the American economy, we needed more power in more places. The most efficient way to generate electricity was by building large, centralized plants, staffed by trained professionals. This approach was consistent with the teachings of the industrial revolution, and solved some of the era's air quality problems by moving major pollution sources away from dense population centers. It also created supply chain efficiencies, since coal could be delivered to just a few point sources as opposed to being widely distributed. Today, changes in our society, infrastructure, technology and scientific understanding are making centralized power generation look like less of a bargain. Here are just a few of the drivers changing our view of centralized energy generation:

DRIVERS FOR DISTRIBUTED GENERATION

Infrastructure

As discussed in Chapter 1, there are serious threats to the energy transmission infrastructure in this country. Energy use is expected to increase slowly but steadily over time.[75] Our existing infrastructure is aging and increasingly unreliable. The American Society of Civil Engineers has given our electricity infrastructure a D+ rating, and found that significant power outages have increased from 76 in 2007 to 307 in 2011.[13] As factors like these meet major weather events such as storms or heat waves, grid reliability will go down and the business case for on-site generation will get stronger and stronger. While on-site generation and microgrids are not immune to weather effects, they do not require the most vulnerable component of the current power grid—transmission and distribution lines.

Technology

Advances in energy generation technologies have not just made combustion of fossil fuels more efficient, they have created smaller, reliable turbines that can operate on the scale of a single building. Developments in renewable energy and battery storage mean that these technologies are finally starting to operate in a cost-competitive manner. While some incentives may still be needed to help renewables overcome the incumbency advantages of fossil fuels, energy generation technology is now a practical choice for many applications. Finally, inverters and power control systems now allow us to operate on-site generation and storage independently of the grid to support grid power quality, and to generate direct power where we need it within a discrete portfolio. These "microgrids" allow end users the flexibility to work with the grid or independently, while leveraging the least-expensive power options available at any given time. All of this means distributed energy has become a cost-competitive approach, and a practical option that in many cases was not available even 10 years ago.

Efficiency

Generating energy in close proximity to end users results

in a more efficient energy system. In the centralized electricity market, we lose 6 percent of the energy we produce during transmission. Generating power close to the end user would eliminate those losses. Similarly, the heat generated from fuel combustion at most large power plants is simply vented into the atmosphere. By generating electricity closer to the end user with a combined heat and power (CHP) system, waste heat can be captured and used to condition air, create steam, or support industrial processes.

Environmental

Recent surveys have found that 97 percent of climate scientists in the world believe that climate-warming trends over the past century are very likely due to human activities.[76, 77] Our use of fossil fuels is at the center of that human impact. Using those fuels more effectively (with on-site CHP, for example) and replacing them entirely with renewable energy will slow (and, someday, hopefully reverse) our deleterious impact on Earth's climate. While there is some ability to deploy these technologies on the larger, centralized grid, they truly shine when deployed close to end users and in some cases, even blended in with our urban infrastructure.

Social Justice

Moving the burning of fossil fuels away from population centers makes sense, but as our country's population has increased, it has become more difficult to truly isolate large scale power plants. The impacts of air, water and land pollution from energy generation are felt in certain locations and by certain populations more than others. Those too poor to live anywhere but near a power plant breathe dirtier air and typically live shorter lives. A study by the American Lung Association found that "Particle pollution from power plants is estimated to kill approximately 13,000 people a year," with people living near the smoke stacks breathing the highest levels of dangerous pollution.[78] Even those not living directly adjacent to power plants may be affected. Prevailing winds blow emissions from Midwestern coal plants over Northeastern states. Overall, the externalities to traditional fossil fuel energy generation are not being paid equally by all members of society.

Market Knowledge

One of the primary limiting factors to the implementation of more distributed generation in the United States has been the custom, non-commoditized nature of the systems. This has made it difficult to demonstrate a clear business case that will be applicable across geographic locations and industry sectors.[79] This has meant more tentative investment by energy companies and communities and a lack of standardized products and business models. This is changing though. The number of examples of successful, long-term on-site generation installations is increasing. Companies offering these products are getting better at scoping them for unique facilities and power needs. Most importantly, professionals like you are getting more and better information about what resources are available and how to evaluate their potential to help you meet your organization's goals.

The developments discussed above and the technologies presented in this chapter are driving infrastructure decisions away from large, centralized power plants and toward a more distributed, more agile, cleaner network of power generation. We believe this shift will completely change the organization of electric utilities in this country, and change the way we interact with energy as a resource. Communities and businesses that embrace this new paradigm early will have an economic advantage over more grid-dependent communities. If the aging grid, heightened storm activity or domestic terrorism create additional significant outages in the power grid, that economic advantage will be significantly amplified.

This chapter will first discuss how to understand your on-site generation needs and begin creating a plan to further develop this area and move your organization further along the spectrum of energy resilience. It will then explore some of the most common on-site energy technologies to provide you with context for evaluating options. Finally, this chapter will discuss microgrids—a combination of infrastructure and information technology that makes independent energy operation possible for local communities and institutions.

EVALUATING YOUR POWER NEEDS

This section will focus on how to understand your community's energy needs and how that understanding can support your energy resilience planning. Determining critical loads in your portfolio and evaluating the cost effectiveness of specific technologies still requires on-site work by a subject matter expert, but this chapter will advise on how to understand your needs and identify the right questions for further study. The process can be organized into five steps. Steps one through three can be done in any order, depending on what information is easiest to access, or what information is needed most urgently:

- Step 1: Benchmark power needs
- Step 2: Understand options and limitations
- Step 3: Choose a level of resilience
- Step 4: Target certain facilities or functions to be sustained by on-site generation or emergency power
- Step 5: Select the right technologies to meet your demand

Step 1: Benchmark Power Needs

Before making major investments in on-site power, examine how your portfolio of assets uses energy. What is the baseload of each building, and what are its peak loads throughout the year? What drives those peak loads? What is the end-use of energy in each facility? Answering these questions will help you understand how much energy you will need for any given combination of buildings. Assuming you are already pursuing efficiency measures to reduce both base load and peak demand, you should be looking at your data to understand how much energy you would need at any given time during a 24-hour cycle to keep each of your facilities running (both at a minimal acceptable level and during typical operation).

Your energy baseline also helps you understand what technologies may be best suited to meet those needs. If space heating represents a large energy use, CHP may be the best solution. If you have critical needs around cooling, photovoltaics and fuel cells may make sense. If lighting is a critical load, fuel cells may be the best alternative. These technologies will be discussed in greater detail in the next section, but for now, remember that understanding your load will help you evaluate the best technologies to deploy. While it makes sense to perform energy audits of your critical facilities to understand power breakdown, Table 6-1a through 6-1c, from EIA's Commercial Building Energy Consumption Survey, provide valuable guidelines for how buildings typically operate in the United States.[80]

Step 2: Understand Options and Limitations

There may be technical, regulatory, financial, and political/social obstacles to consider when deploying new generation capacity within your community. Here is a short (non-exhaustive) list of critical questions to ask in each category before moving forward:

Technical
- Are there physical spaces you can use to install on-site generation technologies? For solar, this may simply mean a field or a rooftop. For a large CHP system, you may need to construct a new building.
- Depending on the proximity and layout of buildings supported by any given system, electricity distribution lines may need to be constructed. If so, what is the technical feasibility of adding this infrastructure?
- What is the structure of the electric distribution grid for your community? Is it possible to create a single point of interconnection with the utility—thereby allowing the community or subset of buildings to disconnect from the grid and operate in island mode?
- What energy sources are available? How much natural gas

On-Site Generation and Microgrids

Table 6-1a: Major Fuel Consumption (Btu) by End Use for All Buildings, 2003[80]

	Total	Total Major Fuel Consumption (trillion Btu)								
		Space Heating	Cooling	Ventilation	Water Heating	Lighting	Cooking	Refrigeration	Office Equipment	Computers	Other
All Buildings	6,523	2,365	516	436	501	1,340	190	381	69	156	569
Building Floorspace (Square Feet)											
1,001 to 5,000	685	213	46	18	49	96	49	138	8	12	56
5,001 to 10,000	563	212	39	18	43	95	37	57	6	10	46
10,001 to 25,000	899	357	57	52	51	184	29	57	10	20	83
25,001 to 50,000	742	281	63	55	60	140	16	37	7	17	66
50,001 to 100,000	913	325	79	78	67	202	17	35	7	20	83
100,001 to 200,000	1,064	399	84	91	81	234	11	30	Q	33	89
200,001 to 500,000	751	286	58	56	69	170	14	10	8	20	61
Over 500,000	906	292	91	67	81	220	18	19	Q	25	85
Principal Building Activity											
Education	820	389	79	83	57	113	8	16	4	32	39
Food Sales	251	36	12	7	4	46	11	119	2	2	11
Food Service	427	71	29	24	67	42	105	70	2	2	16
Health Care	594	223	44	42	95	105	11	8	4	10	51
Inpatient	475	175	35	38	92	76	11	4	2	7	34
Outpatient	119	48	9	4	3	28	Q	4	2	3	17
Lodging	510	113	25	14	160	124	16	12	Q	6	36
Mercantile	1,021	269	110	68	57	308	26	49	8	11	115
Retail (Other Than Mall)	319	107	25	16	5	111	3	22	3	4	24
Enclosed and Strip Malls	702	162	85	51	53	197	24	27	5	8	91
Office	1,134	400	109	63	24	281	4	35	32	74	110
Public Assembly	370	196	38	63	4	27	3	9	Q	Q	26
Public Order and Safety	126	54	10	10	15	18	1	3	Q	2	12
Religious Worship	163	98	11	5	3	17	3	6	1	1	19
Service	312	145	16	24	4	63	Q	9	(*)	3	46
Warehouse and Storage	456	194	14	20	6	132	Q	36	1	5	48
Other	286	138	18	11	4	59	Q	10	2	5	33
Vacant	54	37	2	1	(*)	4	Q	Q	Q	(*)	8

Table 6-1b: Major Fuel Consumption (Btu) by End Use for All Buildings, 2003[80]

	Total Major Fuel Consumption (trillion Btu)									
	Total	Space Heating	Cooling	Ventilation	Water Heating	Lighting	Cooking	Refrigeration	Office Equipment	Computers	Other
Year Constructed										
Before 1920	303	181	7	11	17	35	17	Q	2	4	15
1920 to 1945	631	318	26	30	43	92	20	26	3	8	64
1946 to 1959	588	284	32	36	46	94	13	27	4	11	41
1960 to 1969	791	353	48	53	68	127	15	42	7	19	60
1970 to 1979	1,191	397	97	86	101	265	32	64	14	29	106
1980 to 1989	1,247	359	122	82	103	298	33	74	17	39	120
1990 to 1999	1,262	352	129	101	84	294	40	91	18	36	117
2000 to 2003	511	122	55	37	39	136	20	41	4	10	47
Census Region and Division										
Northeast	1,396	674	55	76	93	239	38	63	13	32	113
New England	345	186	10	15	20	55	Q	21	2	7	21
Middle Atlantic	1,052	488	45	60	74	184	31	42	11	25	92
Midwest	1,799	874	67	109	107	313	38	92	16	36	146
East North Central	1,343	676	46	83	77	230	28	62	12	27	102
West North Central	456	198	21	26	30	83	10	30	4	9	45
South	2,265	519	308	179	191	543	82	167	20	58	198
South Atlantic	1,241	278	162	99	100	311	40	95	12	37	106
East South Central	340	113	27	25	32	72	10	25	2	6	28
West South Central	684	128	119	54	59	160	31	47	6	15	64
All Buildings	6,523	2,365	516	436	501	1,340	190	381	69	156	569
West	1,063	298	86	72	109	245	33	59	20	30	112
Mountain	446	167	31	27	41	94	7	20	Q	9	43
Pacific	617	131	55	45	68	151	25	39	Q	21	69
All Buildings	6,523	2,365	516	436	501	1,340	190	381	69	156	569

On-Site Generation and Microgrids

Table 6-1c: Major Fuel Consumption (Btu) by End Use for All Buildings, 2003[80]

	Total	Space Heating	Cooling	Ventilation	Water Heating	Lighting	Cooking	Refrigeration	Office Equipment	Computers	Other
Climate Zone: 30-Year Average											
Under 2,000 CDD and –											
More than 7,000 HDD	1,086	555	27	66	67	180	22	62	9	18	80
5,500-7,000 HDD	1,929	913	75	109	127	352	46	94	17	41	154
4,000-5,499 HDD	1,243	463	78	77	95	259	37	65	19	32	118
Fewer than 4,000 HDD	1,386	328	143	108	134	330	53	101	15	35	139
2,000 CDD or More and –											
Fewer than 4,000 HDD	879	106	193	76	78	219	32	59	8	29	79
Number of Establishments											
One	4,167	1,562	289	272	362	790	139	290	42	87	334
2 to 5	1,161	460	84	74	75	245	27	55	13	26	101
6 to 10	378	135	32	25	19	89	8	11	4	18	37
11 to 20	307	82	37	25	17	82	7	11	3	8	34
More than 20	473	98	74	39	27	131	9	13	7	17	57
Currently Unoccupied	37	28	Q	(*)	Q	3	Q	Q	Q	Q	5
Energy Sources (more than one may apply)											
Electricity	6,522	2,364	516	436	501	1,340	190	381	69	156	569
Natural Gas	5,042	1,878	356	333	436	987	183	259	47	114	448
Fuel Oil	1,867	683	142	129	204	389	36	47	25	52	160
District Heat	1,029	609	34	70	60	150	Q	12	5	29	50
Energy End Uses (more than one may apply)											
Buildings with Space Heating	6,370	2,365	488	425	490						
Buildings with Cooling	6,149	2,140	516	420	479						
Buildings with Water Heating	6,158	2,179	493	417	501	1,297	183	361	68	153	540

See "Guide to the Tables" or "Glossary" for further explanations of the terms used in this table. Both can be accessed from the CBECS web site - http://www.eia.doe.gov/emeu/cbecs.
(*)=Value rounds to zero in the units displayed.
Q=Data withheld because fewer than 20 buildings were sampled for any cell, or because the Relative Standard Error (RSE) was greater than 50 percent for a cell in the "Total" column.
Note: Due to rounding, data may not sum to totals.
Source: Energy Information Administration, Office of Energy Markets and End Use, Form EIA-871A, C, and E of the 2003 Commercial Buildings Energy Consumption Survey.

flow could be delivered given current pipe specifications? Are other fuel sources (such as methane or agricultural waste) available nearby?

Regulatory
- Is there a net-metering law in your state that would allow you to sell excess power back into the electric grid? If so, what are the system size parameters for participation?
- Is your local utility supportive of your plans to implement on-site generation? Are there ways your system could support the stability and cost-effectiveness of the local grid at large?
- Are there regulatory hurdles to installing additional energy distribution lines between buildings?
- How much generation is possible in your location before you will be subject to regulations pertaining to commercial generation facilities?

Financial
- How will the cost of delivered power from new generation systems compare to current, utility-provided electricity?
- How long is the expected payback on these new investments, and is that payback within acceptable parameters for the community? (Try to include abated risk in your payback calculations.) How much would future power outages cost your institution?
- What are the prices of natural gas and other fuels and how do these prices compare to the current price of electricity (per Btu)?
- What financing mechanisms are available to pay for large projects like this, and what are their relative advantages for this project?
- Do you want to own this system or does it make sense to contract a third party to build and own the system, signing

a long-term power purchase agreement with that third party for the generated energy?

Political/Social
- Is there political will in the community to increase resilience and add on-site generation?
- Are there stakeholders or interest groups that may oppose this effort?
- Will multi-year financing efforts be subject to changes in political leadership and expose the project to risk of defunding?
- Are there local firms that can provide some or all of the expertise required to design, build, operate, and maintain these systems? If not, is there an opportunity to attract new jobs to your community through this kind of investment?

Answering these questions is far from simple. However, understanding these issues is critical to understanding the environment in which your projects can proceed. Some closed doors can be opened, but others may represent permanent barriers. In those cases, you have saved yourself the trouble of pursuing an untenable solution. As the adage says, "don't let the 'perfect' become the enemy of the 'good.' A general understanding of these questions will be sufficient to begin designing your solutions. New obstacles will arise as you move forward, and a requirement of absolute certainty may leave you in an endless spiral of research and analysis, and deplete the enthusiasm of project participants.

Step 3: Choose a Level of Resilience

Setting a target level of energy resilience may be the most important step you take in this process. It will create parameters around your entire effort. This step could easily come first or second—depending on whether you want to let an understanding of your portfolio drive this decision, or let this decision drive your portfolio evaluation. As with most complex undertakings, the safe approach is usually to aim low to reduce the complexity of the project and create some early success you can build on. For some communities, the potential costs of grid disruption are simply too high to settle for incremental progress. Review the energy program maturity model in Chapter 4 for a more in-depth discussion of levels of energy resilience management.

The following is a simple, three-tiered way to think about your level of energy resilience in terms of on-site generation:

Bare Bones

You identify a few critical functions/facilities needed in the event of a long-term grid disruption. These may include supplying power to first responders, emergency coordination staff, and possibly a central shelter location. The generation assets may be a mix of on-site power tied to discrete functions and emergency backup power from batteries or small generators.

Reduced Load

You want to be able to continue operations during a grid event, but need to do so with a significantly reduced load profile. You deploy a microgrid to supply power and heat to much of your portfolio. When operating in island mode, you implement a load-shedding protocol to shut down non-essential buildings and certain uses with critical facilities.

Full Operation

You implement a robust microgrid to control generation and storage capabilities to meet your community's typical load profile. During normal operation, you may sell some electricity back to the grid or provide some level of load regulation for the electric utility. During a grid disruption, you operate at full power but in island mode. See the sidebar on Net-Zero Energy Communities for an example of how the National Renewable Energy Laboratory proposes such a solution using renewables.

In all three of these scenarios, your on-site generation systems can be operating year-round (or when cost effective) to deliver power in tandem with the grid. The right solution likely depends on the answers you found in Step 22. If you can generate power more cost effectively on-site than you are receiving from the power grid, it may make sense to deploy a full operation model. If there are significant barriers to on-site generation, you may need to settle for a bare bones approach, and try to maximize that system with excellent energy efficiency and emergency planning efforts.

Net-zero Energy Communities

The concept of net-zero buildings has been around for several years, referring to buildings that can produce as much energy as they use via renewable energy technologies. Increases in renewable efficiency and the commercialization of microgrid control technology means this concept can now be expanded to the community level. The concept of a Net-zero Energy Community is outlined in detail in a report from the National Renewable Energy Laboratory.[81] This report details the concept and describes valuable strategies for developing such a project in intermediate steps to ensure that environmental goals remain in balance with budget realities.

The NREL report provides valuable guidance for setting small goals to make the ambitious goal of net-zero seem more manageable. "The intermediate steps can be anything that community members can visualize and can be large or small. For example, 'the community has verified that it has generated enough power from renewable sources to claim it is net-zero for the equivalent of the elementary school or all buildings in the community.' One set of intermediate targets could include:
- The equivalent energy offset to accomplish all the "Option 0" strategy in the renewable energy supply hierarchy. The Option 0 case represents the amount of energy (or carbon) needed to power the community under a scenario where all energy-efficiency measures and design improvements are included to maximize cost-effective energy savings.
- The equivalent energy to offset the energy used by buildings in the community.

> - The equivalent energy to offset the energy used by transportation in the community.
> - The equivalent energy to offset all other uses of energy in the community.
> - The equivalent energy to offset the added energy growth that will come over time as the community grows."

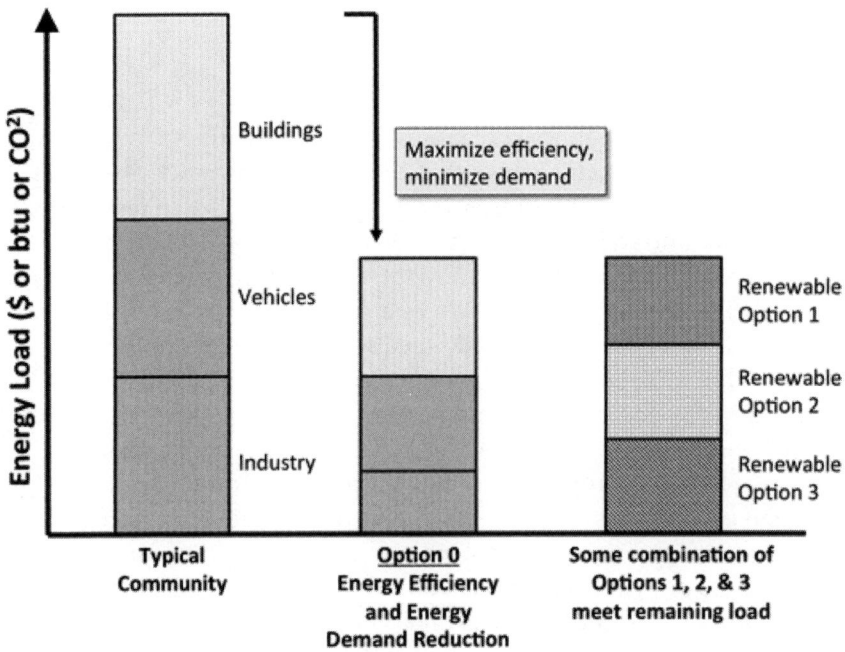

Figure 6-1: The Net-zero Concept.

Step 4: Target Critical Facilities or Functions

Once you understand your level of resilience, you will know how much energy will be available if there is a grid power outage. Power will suddenly become a finite resource and careful thought must be given to how that resource will be allocated. These decisions cannot be made after the fact. The placement of generators and energy storage, the transmission capabilities of a microgrid, and the energy component of emergency planning all require clear decisions about how power will be distributed should your community go into island mode. These will include selecting buildings that are critical to keep operational in the event of a grid outage. It may involve selecting specific systems within those buildings to prioritize. For example, a laboratory or kitchen may prioritize refrigeration over lighting, whereas law enforcement may prioritize telecommunications equipment above all else.

To start this process, here are the three priorities that should be used to evaluate where power should be directed when entering an energy-limited event. They should be addressed in this order.

> **Priority 1:** What is needed to ensure the immediate safety of the people in this community? A long-term power outage is likely to be coupled with some other event (most commonly a storm or natural disaster). Who does your community rely on when these things happen? What do they need to do their job? Who to those people, in turn rely on? Think through the logistical realities of this so that you can not only keep the lights on at the headquarters of your first responders, you can ensure that those people can get where they need to go and have access to meals and support personnel. Reading your organization's existing emergency preparedness plans will clue you in to all the critical functions that should be given energy priority.
>
> **Priority 2:** What power is needed to protect your ability to accomplish your core mission? Here we are not just talking about continuing business as usual. We are talking about the minimum power level to prevent irreparable damage to your organization's health. For a hospital, this may not mean the ability to continue to treat patients, but the ability to maintain the stability of patients as they are transferred to another facility. It may also mean protecting the environmental parameters of critical research or stored blood. For a corporation, this may mean keeping customer-facing IT in-

frastructure running, even if all other aspects of the business need to be suspended. This priority is about avoiding damage that cannot be later repaired—whether that is damage to infrastructure, research, or even reputation.

Priority 3: How can you support a speedy return to normal operation when grid power is restored? U.S. utilities have dedicated and capable engineers and repair personal that work long hours in dangerous conditions to restore power when it goes out. When they succeed, many organizations are still not able to quickly return to normal operations because they have not made a plan for ramping back up. For this priority, think about what will be needed to engage with members of your community—to keep them apprised of developments and prepare them for changes. Think about what damage may have been caused by the power outage that will need to be addressed before normal operations resume. Think about the systems that will need to be tested to ensure safety or recalibrated to ensure accuracy. After addressing priorities 1 and 2, dedicate additional power to support the operational staff that will manage and effect this transition back to normality.

The discussion above addresses priorities in a power outage. As you know, there are other threats to energy resilience such as price spikes. When energy costs rise dramatically, a similar analysis must be done. You are not thinking about 'what do we absolutely need?' and instead 'what can we probably do without?' Understanding elements of non-essential operations that can be shut down in the face of price spikes, brown outs, or highly lucrative demand response programs is an excellent way to manage energy costs and prepare to support grid-wide resilience.

Step 5: Select the Right Technologies to Meet Demand

Once you understand your desired level of resilience, your energy load patterns, and the limitations and opportunities facing

On-Site Generation and Microgrids 147

this project, you can finally ask the question, "So, what should we install?" All distributed-energy resources were not created equal, and it is important to select the technologies that best suit your specific goals. The next section goes into some depth on the major distributed-energy technologies in use today and where each of them makes the most sense. You will need to match up what you learned and decided in Steps 1-3 with technical advice on how each technology can support your system's energy resilience. Here are some specific considerations to keep in mind as you evaluate technologies:

Interconnectivity

Your ability to interconnect multiple generation sources and loads is critical when it comes to certain energy resilience strategies. If you cannot create a local microgrid that can operate in island mode, you have clear limitations around how distributed-energy sources can be used. In that case, each generation source would need to power a discrete building and have separate wiring within that facility to function in case of grid disruption. If you can interconnect loads and generation sources in a microgrid, then you can produce power anywhere within that microgrid to serve any connected load.

Scale

The amount of power you are trying to produce will significantly impact your generation choices. If small amounts of power are needed, a photovoltaic system (possibly with battery storage) may be the right answer. If you need large amounts of base load power, a gas-fired turbine in a CHP application may be more appropriate.

Desired Resilience

The kinds of power sources you use also depend on the kinds of threats you face and how protected you want to be. We typically think of generators or fuel cells that use natural gas to be highly resilient because natural gas pipelines are underground and not susceptible to transmission outages the way electric transmission lines are. However, natural gas is not immune to price spikes, fuel shortages, or even delivery interruption. Communities that feel natural gas delivery is not reliable enough will likely want a technical strategy that employs more renewable energy such as solar, wind, biomass, and geothermal. The most resilient system will employ multiple generation types to diversify supply and reduce risk from any one kind of system failure.

Load Types

The way your facilities use power also drives your technology choices. If demand is highly variable, with dramatic spikes during certain times of day, a battery storage system may be ideal to flatten out the load profile. If heating is a critical element of the base load, or more to the point, the emergency load during a grid disruption, then a CHP application that provides heat as a byproduct of power generation may make the most sense. If you have more interruptible loads that can be shut off for short periods as needed, you should consider how a microgrid design could support participation in demand-response and load-shedding programs—adding an additional revenue stream to the system.

Available Resources

Geography is clearly a driver when it comes choosing renewable resources, due to the availability of sun, wind, and other natural features that fuel renewable generation. We can also think about political geography as a driver, since each state has slightly different regulatory and incentive structures. You should also consider your access to other fuel types. The questions from Step 2 about access to natural gas and other fuel feed-stocks will help inform your decisions about which technologies are possible and which of those will be the most cost-effective and resilient.

As you walk through these steps, there are best practices to pursue—and planning pitfalls to avoid. Below are some of the most critical in the context of distributed generation.

Best Management Practices

Think of the entire project as one integrated system. Whether you are going to deploy a robust microgrid or a series of small generation projects, they should be thought of as a single system. Remember, your purpose is to improve energy resilience, and all parts of this project should be considered components in your energy resilience machine. Using systems thinking will also help you understand how each component can work in concert with other components, making your system more than the sum of its parts.

Second, use every knowledge resource available to you. The National Renewable Energy Laboratory is working with communities all over the United States to help them deploy cutting-edge technologies. Other communities (local governments or colleges or companies or military bases, for example) may be willing to talk to you about their experiences and share lessons learned about their processes. (The case studies in this book are just a sampling of the expertise available across industries.) Finally, you may have expertise in this area within your own community (e.g., residents or business leaders or staffers).

Finally, develop an infrastructure implementation plan with immediate, tangible benefits to the community that is also scalable for later, more comprehensive efforts. Understand that an entire microgrid will not be built and operational within a week. Attempt to schedule new resources to come on-line as they are completed, and make them immediately available to the community. This will create early, visible success that can generate continued momentum for project implementation. At the same time, whatever level of resilience you pursue, think about how this project, if successful, might be expanded in the future. Giving this some thought early in the process may lead to design decisions that make future projects much more successful.

Pitfalls to Avoid

It is important to complete Steps 1-3 before considering which technologies to implement. A team member excited about the idea of fuel cells or CHP may try to make this effort about that specific technology. Instead, keep your focus on improving the energy resilience of your community, and do not be distracted by any specific technology early in the planning process.

Do not make policy or technical decisions based on one person's opinion or previous experience with a technology—unless that person is a subject matter expert that the team has decided to trust in those matters. Distributed generation technologies are a rapidly developing industry. If one person has it in his or her head that "solar is too expensive," or "large batteries aren't reliable," their understanding of these technologies may be limited or dated. Get the facts before selecting or rejecting any specific solutions.

Finally, do not take your technical advice (solely) from a product manufacturer or anyone with a vested interest in which technology you ultimately employ. Product manufacturers can be valuable allies in these efforts because they understand better than anyone how to deploy their systems to the maximum possible benefit. Remember though, their job is to sell those systems. Balance their advice with advice from technology-agnostic experts who have a deep, cross-cutting knowledge of distributed-energy systems but will not benefit from any specific choices you make. These professionals can typically work with multiple product vendors to leverage their in-depth knowledge without coloring the final recommendations to your team.

When you do reach the point where you are ready to select specific technologies for deployment, you will need to conduct considerable research and evaluation. An overview of some of the most prominent on-site generation and storage technologies is provided in the following section to help frame your thinking around these decisions.

TECHNOLOGY OVERVIEWS

On-site Generation Technologies

Until recently, most on-site generation consisted of diesel-burning back-up generators—typically providing emergency backup power for a few hours or as much as a few days. When not being used in an emergency, these generators sat idle alongside large fuel tanks, taking up space and generating no revenue source for their owners. Improvements in on-site generation technologies and mounting concerns around energy resilience have driven an increase in the options available for deploying these systems. It is now cost effective in certain cases to run these systems year-round—creating the economic and resilience benefits discussed above. However, without a local network for distributed generation, these systems must go off-line during a grid outage. If they did not go off-line, they would be sending electricity back into transmission lines during a grid outage, where it could seriously endanger repair crews. The solution is to develop microgrids—a technology described in detail below.

It's helpful to first see the broad landscape: the most common distributed-energy technologies being used today. This book cannot present an exhaustive list of every energy system. Some technologies like geothermal heat pumps and solar thermal, for example, are not addressed because, though they can contribute to resilience, they do not produce electricity. Other technologies such as hydro turbines and geothermal electricity production are not addressed because they tend to only make sense at the utility scale or in special locations.

This section will likely be remedial for energy professionals with a strong working understanding of renewable energy technologies. The descriptions below are brief summaries for readers who may only need a high-level understanding of technology options.

Thermoelectric Turbines

The thermoelectric turbines in use in distributed systems employ essentially the same technology used in large power plants, scaled down for use in a building or microgrid. They use a combustible fuel source and produce consistent, baseload power any time it is needed. The most common fuel source for these turbines is natural gas. However, other fuels can be used such as diesel, gasoline, methane (from landfill collectors), or biogas (generated via farming products put through a digester process). Methane and biogas may require more engineering to harness but represent a more environmentally friendly and sustainable fuel source.

These systems are dependent on the regular delivery of fuel, which limits their resilience. However, natural gas lines tend to be underground and are therefore much less vulnerable to damage from inclement weather. Our nation's natural gas infrastructure is aging, but the underground distribution pipes are well protected, making the natural gas system generally more resistant to system shocks.—particularly extreme weather.

How It Works

First a compressor draws air into the engine and pressurizes it. This highly pressurized air then enters the combustion chamber. In the chamber, fuel injectors fire the fuel source in to mix with the air. As this mixture is ignited, it boils water to create a high-temperature, high-pressure steam that is sent into the turbine section. This steam picks up speed as it expands in the turbine section, rotating the blades of the turbine as it passes over them. Rotation from the turbine blades is used to generate power as well as to drive the compressor.

An improvement on the traditional generator is the combined-cycle approach. In this modification, the waste heat from the combustion process is used in two ways. First, a heat recovery steam generator captures waste heat to turn water into steam, turning a second turbine. The remaining heat in the system after that steam generation is used to preheat the air entering the compressor at the head of the system. This approach generates more energy from the same quantity of gas and increases turbine efficiency from around 20-35 percent to an impressive 60 percent or more.[82] See Figure 6-2.

Where it Works Best

Thermoelectric turbines can work well in a variety of locations. They are a proven technology, and can typically generate

On-Site Generation and Microgrids

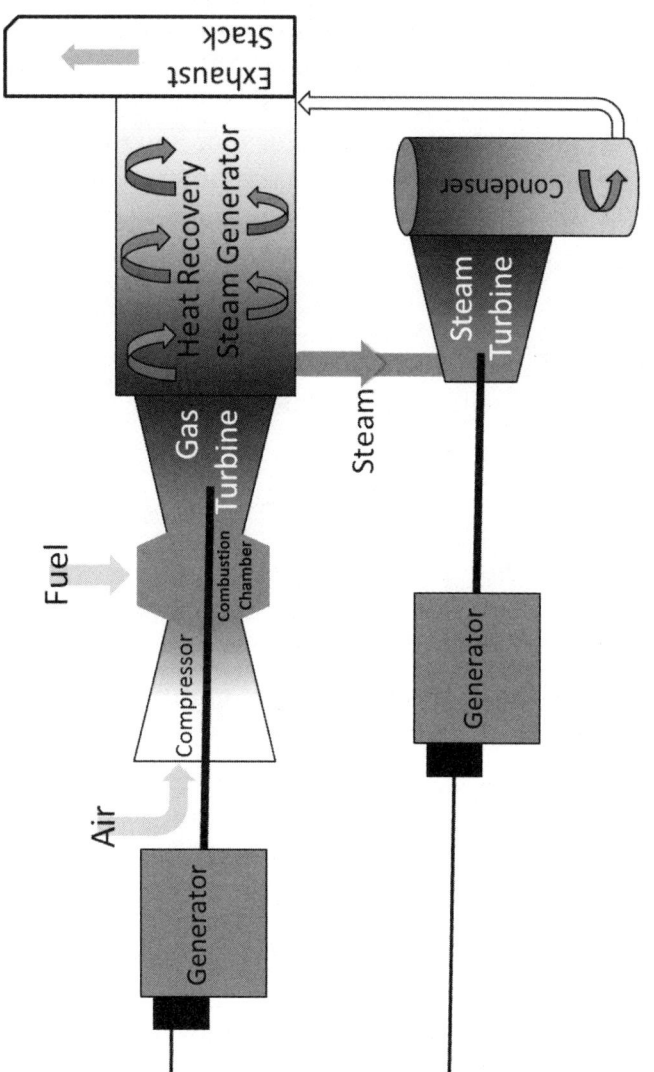

Figure 6-2: Schematic of a combined cycle thermoelectric turbine.

much larger amounts of baseload power than other distributed generation technologies. However, they do require an external fuel source. From a resilience perspective, these systems work best in places with hardened natural gas supply. This means places where all gas distribution lines are underground and the natural gas refinery feeding your distribution system is in an area with low risk for events such as storms, flooding, earthquakes, or wildfires. Alternately, if you have local access to methane from landfill gases, or biogas from local farming operations, making use of them could make your turbine system both more resilient and more sustainable.

Pros:
- Can provide large amounts of baseload power
- Considered a mature, reliable technology
- Gas delivery infrastructure is more resilient than the electrical grid

Cons:
- Require an external fuel source (gas lines are not invulnerable), reducing the resilience factor
- Turbine systems are complex and highly technical, and require trained staff to maintain
- Associated with high initial costs and are ideal for larger applications
- Require a dedicated space to operate and may require constructing a new facility to house
- While efficient, these systems still use non-renewable fossil fuels, produce air pollution and contribute to global climate change

Fuel Cells

Fuel cells deliver constant, on-site power generation that can serve base loads while producing no harmful emissions. This technology uses hydrogen as a fuel with water as its only byproduct. The fuel cell's straightforward approach makes it incredibly efficient. Noriko Behling explains this clearly: "Fuel cells are singularly remarkable in their potential for efficiently converting the energy locked up in chemical bonds to electrical energy. This efficiency is achieved because fuel cells convert the chemical energy contained in a fuel into electrical energy in a single step, extracting more useful energy from the same amount of fuel than any other know device."[83]

There are several factors slowing the growth of fuel cells. First, the newness of the technology for commercial applications means that many decision-makers see fuel cells as an exotic or experimental solution. In fact, fuel cells have been operating effectively as on-site generation for many years. Cost and maintenance of these systems is a valid concern. The current designs use rare and expensive elements, keeping fuel cells from being cost competitive with other energy sources. Hydrogen is relatively new as a fuel. We do not have long experience with it nor the infrastructure that we have for fuels such as natural gas or propane. Finally, hydrogen cannot truly be mined or pumped out of the ground. It has to be produced using energy-intensive chemical processes. This means the efficiency and sustainability of fuel cells is determined by the efficiency and sustainability of the hydrogen production process. Some industry experts bemoan the slow pace of development for this technology by insisting, "Fuel cells are the power source of the future ... and always will be."

To combat some of these barriers, fuel cell manufacturers are developing products and production methods to improve fuel cell marketability for commercial applications. Many fuel cells now include technology to reform fuels like natural gas and methane into pure hydrogen in the fuel cell unit itself. This process does create additional waste byproducts beyond water (such as CO^2), but much less than would be created by burning those fuels to generate energy in a turbine. More importantly, this feature creates fuel flexibility and makes fuel cells a more practical option for distributed generation. Industry players are experimenting with which advanced materials to use as catalysts and in the membrane layer to maximize efficiency, and they are replacing metal parts (a weak point for lifespan and reliability) with ceramics and other materials.[84]

How It Works

A fuel cell uses an electrochemical energy conversion process to turn hydrogen and oxygen into water while also generating electricity. In very simple terms, it does this by introducing a chemical catalyst to break electrons off the hydrogen and generate electricity. The now-ionized hydrogen atoms then join with oxygen atoms to form water. A single fuel cell produces a small amount of energy so a number of them are stacked together to scale up the system to the desired output. There are different kinds of fuel cell technologies available today. Currently, the best technology for building or grid-scale applications is the solid oxide fuel cell. This technology is maturing rapidly and new products are on the horizon. See Figure 6-3.

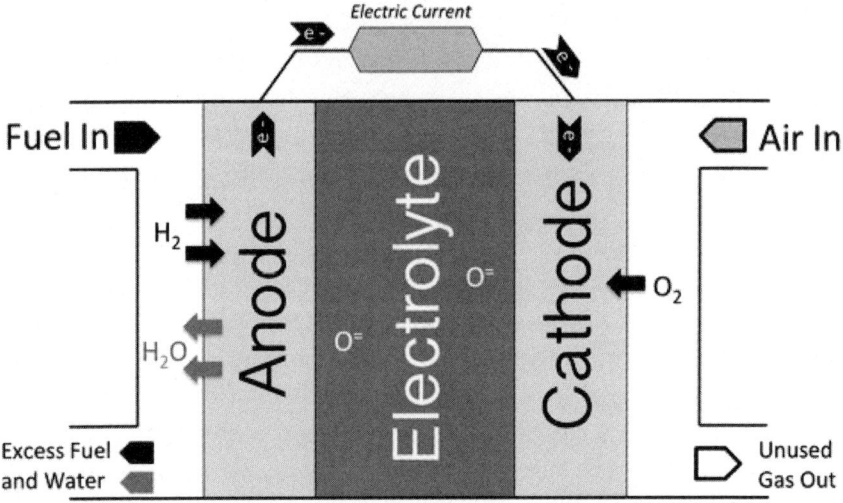

Figure 6-3: Schematic of a Fuel Cell

On-Site Generation and Microgrids 157

Where It Works Best

Fuel cells work best in applications where baseload power is needed but there are concerns around emissions (environmental sustainability), footprint (not much space), or noise (assets will be situated near occupied areas). Some form of fuel is needed so these systems work best in areas with lower costs of hydrogen or more common fuels like natural gas. Fuel cells can be installed with a CHP system so they work well where there is a thermal load. However, if generating heat is a primary goal, a fuel cell will generate much less than a thermoelectric turbine. Although fuel cells can be run at partial load, doing so results in less generation and therefore a longer payback from the system. For this reason, fuel cells are typically run at full capacity and should not be oversized beyond the requirements of the site.[85] As there are certain fixed costs for fuel cell installations, larger systems will tend to have a lower installed cost per kilowatt. Currently, the states with the strongest support of fuel cells applications include California, Connecticut, Delaware, New Jersey, New York, South Carolina and Texas.[86]

Pros:
- Provides emissions-free base load power
- Highly scalable technology
- Small footprint and clean, quiet operation allow for deployment in even dense urban environments
- Hydrogen is the most common element in the universe
- Hydrogen for fuel cells can be produced using a variety of different methods and fuels

Cons:
- High capital costs
- High maintenance costs
- Systems do not work well in extreme heat or cold

- There is little hydrogen infrastructure in the United States
- Pure hydrogen must be produced through a chemical process

Combined Heat and Power

The goal in any exothermic power generation process is to use every bit of energy you can from the initial fuel and dump as little waste heat (energy) out into the air as possible. Both combustion turbines and fuel cells create waste heat as part of their energy generation process. Fuel cells are much more efficient, creating considerably less waste heat per unit of energy produced. However both fuel cells and combustion turbines can benefit from additional use of waste heat. As discussed above, combined cycle turbine systems use waste heat from the combustion process to increase system efficiency. CHP takes this process further by using the remaining heat in the system (after the combined cycle) to enable, or increase the efficiency of, other applications. The most common use of this heat is to warm air to create thermal comfort in nearby buildings. Using advanced absorption chillers, this heat can also be used to create air conditioning. In manufacturing settings, this heat may be used to increase the efficiency of an industrial process. CHP is a logical approach for distributed generation because the waste heat generated needs to be near energy end-users who can make immediate use of that steam or hot air. CHP is a critical technology for any community installing gas turbine generation, and it should be considered for fuel cell applications as well.

Used correctly, CHP can reduce energy costs for end users while reducing the environmental impact of energy use. According to the U.S. EPA and DOE, "installing an additional 40 GW of CHP (about 50 percent more than the current levels of U.S. CHP capacity) would save approximately one Quadrillion BTUs (Quad) of energy annually and eliminate over 150 million metric tons of CO emissions each year (equivalent to the emissions of over 25 million cars). The additional CHP capacity would save energy users $10 billion a year relative to their existing energy sources. Achieving this goal would also result in $40-80 billion in new capital investment in manufacturing and other U.S. facilities over the next decade."[87] At the same time, increasing use of CHP would reduce the need for central generation and wasteful transmission, reducing power sector investments by hundreds of billions of dollars globally over the next 20 years.[88]

How It Works

CHP can be thought of as an additional cycle in the combined cycle turbine approach described earlier. After turning generation turbines, the hot water and steam that remains is run through a heat exchanger that can take the useful heat remaining and use it for space heating or industrial processes. The hot water or steam can also be pumped through insulated pipes to nearby buildings for similar uses there. See Figure 6-4.

Where It Works Best

CHP works best where there is a thermal load that can be met using the system—particularly in colder climate areas. The earliest systems focused on industrial applications where process heating requirements were high, but the technology has been proven to be effective in commercial and residential locations as well. CHP works best in these settings where it can feed heat to a district heating system that can distribute heat and steam to multiple buildings. The International Energy Agency presented the following as criteria for a location where CHP would work well:[88]

- A ratio of electricity to fuel costs of at least 2.5:1;

- Relatively high requirements for heating and/or cooling (e.g. annual demand for at least 5,000 hours);

- The ability to connect to the grid (if present) at a reasonable price with the availability of backup and top-up power at reasonable and predictable prices; and

- Availability of space for the equipment and (for non-DHC related systems) short distances for heat transport.

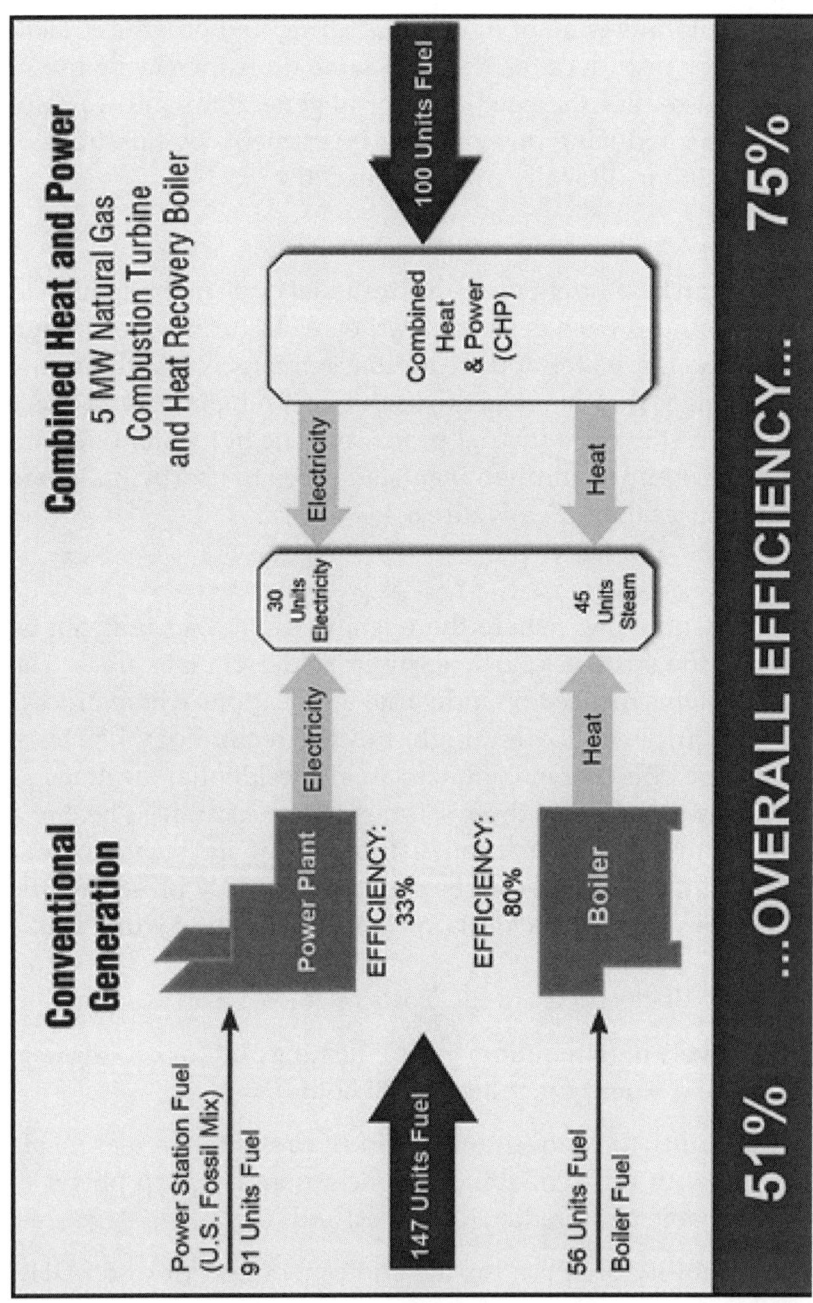

Figure 6-4: Efficiency Gains of CHP: one example.[89]

Pros:
- All the pros of combined-cycle gas turbines or fuel cells but with increased system efficiency
- Provides heat as well as power, further contributing to the resilience of institutions using these systems
- Can be purchased as pre-packaged system to reduce project complexity

Cons:
- All of the cons of combined-cycle gas turbines or fuel cells with slightly more cost and complexity
- Requires infrastructure to transport hot water or steam between buildings to be effective across a portfolio
- Creating cooling with these systems requires expensive absorption chiller technology

Photovoltaics

Creating energy from sunlight is an excellent idea. Sunlight is free and ubiquitous. However, it has taken a considerable amount of time for the technology of photovoltaics (PV) to become cheap and efficient enough to serve as a cost-effective option for power generation. The good news is that the cost of this technology has been dropping rapidly while public acceptance of solar panels integrated into our building stock has been increasing. According to the Lawrence Berkley National Laboratory, in 2012 alone, the installed cost of PV systems dropped 6 to 14 percent from the prior year (depending on the size of the system).[90] The same report stated that the median installed cost of PV systems for residential or small commercial systems (<10kW) was $5.30 per watt while the cost of larger commercial systems (>100kW) was $4.60 per watt. See Figure 6-5.

As costs are dropping, system efficiencies and reliability are increasing. Of course, PV produces no energy at night and produces less when cloud cover blocks the sun. This limitation (typ-

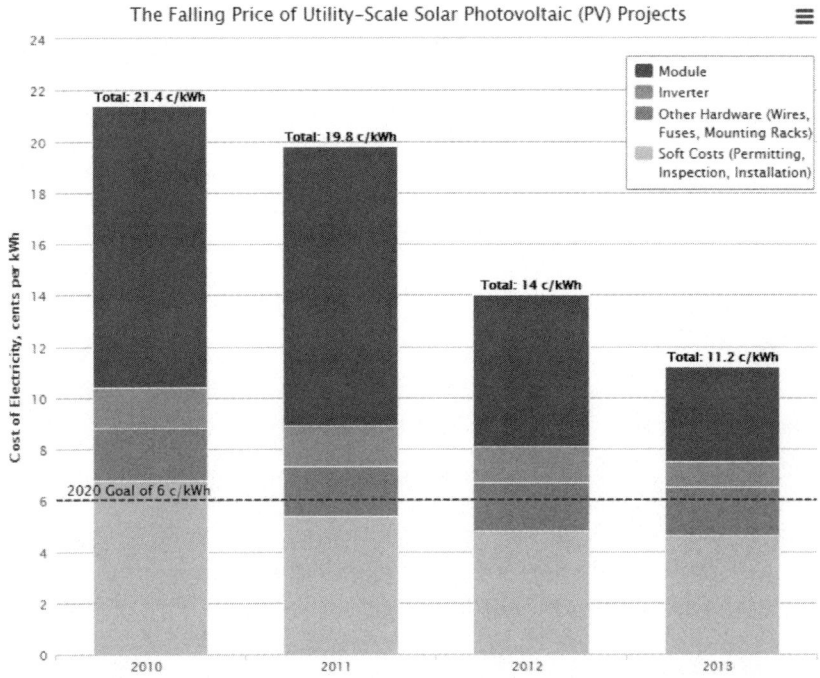

Figure 6-5: The Falling Price of Utility-Scale Solar Photovoltaic Projects

ically referred to as "intermittent power") will always limit PV's ability to play a major role in power production. However, when PV is coupled with technologies that store energy, like batteries or fuel cells, its potential for effecting energy resilience is dramatically increased. These are expensive technologies and their inclusion in a system may price solar out of cost competitiveness with traditional fossil fuels. For now, solar is a cost-competitive addition to any energy system but it will remain a niche player until energy storage technologies also move down the cost curve.

How It Works

A PV cell is made up of semiconductors that absorb the energy from photons in sunlight. The photons strike the atoms in the semiconductor, freeing electrons. These free electrons are then directed by an electrical field to travel in one direction, creating current.

Where It Works Best

Photovoltaics work best where there is abundant sunshine, high electricity costs, and financial incentives for new installations. PV also is popular where there are "net metering" laws, which are grid interconnection regulations requiring utilities to pay for solar power generated by an end user and delivered to the grid. It is rare that all of these factors are in play in one location, but the technology has advanced to the point that just one or two of these factors will create a situation in which the deployment of PV is financially and technologically practical.

In 2014, the Environment America Research and Policy Center identified 10 states that have been clear leaders in using financial and regulatory means, along with outreach to residents and businesses, to promote new deployment of PV systems. Those states are Arizona, California, Colorado, Delaware, Hawaii, Massachusetts, Nevada, New Jersey, New Mexico, and North Carolina.[91] Honorable mentions were made of New York, Vermont, and Georgia—states with fast-growing solar markets and strong new solar policies or programs implemented since mid-2013. Not all of those states have outstanding access to sun. Not all of them have extremely high energy prices. The parameters where solar power works well are clearly widening.

Pros:
- Emissions-free power source
- Zero fuel costs
- If connected to the right system and well maintained, PV will produce electricity throughout almost any disaster
- Highly scalable
- Can be installed in a variety of locations (including building rooftops)
- Low operations and maintenance costs

Cons:
- Intermittent power source requiring an energy-storage technology to sustainably meet base loads

- Large areas of land (or rooftop) is required to generate significant amounts of energy
- PV panels produce direct current and require inverters to provide alternating current power

Wind Power

After hydro power, wind power (in the form of a windmill) is probably the oldest energy generation technology. However, in their role as a modern electricity source, wind turbines have only recently become a cost-effective option. Total wind capacity in the United States has increased roughly 900 percent since 2003 and installation of small wind turbines (applicable for distributed-energy applications) has increased almost 500 percent in that time.[92] In 2012, wind power represented the largest non-hydro renewable energy source and the largest Source of U.S. electric-generating capacity additions. The impressive increase in wind power in the United States has been largely attributed to state-level renewable portfolio standards (RPS)—requiring utilities to incorporate a certain amount of renewable energy in their portfolio—and the national Production Tax Credit.[93, 94] The PTC is credited with contributing to the expansion of wind power capacity in the United States but it has also suffered from short approval timeframes. This created a boom/bust cycle of wind development where developers ceased any new operations toward the end of each PTC authorization cycle—waiting to see if the PTC would be renewed before committing to new projects.[95] The PTC expired at the end of 2013 and, as of now, has not been renewed. The current challenge with wind power is less about its technical applicability and more about concerns around impacts on natural beauty, a.k.a. "view shed," and wildlife preservation—particularly as off-shore wind developments are being proposed.

How It Works

Wind turbines are, at a high level, quite simple. Wind turns the blades which turn a shaft connected to a generator. This creates direct-current power that must go through an inverter to provide alternating-current power. As a rule of thumb, the power generated by a turbine is roughly proportional to the square of its blade length. Generally speaking, doubling the length of the blades will quadruple the power output of the turbine. That relationship does not seem to bode well for the smaller wind turbines you would apply in a community or campus setting, but there is still a compelling case in certain locations. Tradition turbines with propeller-like blades are now being supplemented with vertical-axis turbines—some of which are much more applicable to urban environments.

Where It Works Best

Wind turbines work best where there are strong, consistent winds, high electricity costs, financial incentives for new installations, and net-metering laws. At the community scale, we should think about locating turbines away from large buildings—in large open fields or on ridgelines. For denser urban environments, architects and engineers have been increasingly finding ways to use small turbines to harness air flows created by buildings and even passing traffic.

Pros:
- Emissions-free power source
- Zero fuel costs
- Turbines take up very little space and can easily share usage of the land on which they're sited

Cons:
- Intermittent power source requiring an energy storage technology in order to substantially address base loads
- Need to be turned off in extremely high wind and can be damaged by extreme storms
- Extremely visible and can create noise and visual distraction from natural environment—characteristics that nearby residents may find objectionable

- There is a lack of public understanding around how wind turbines affect bird and bat populations
- Wind turbines produce direct current and require inverters to provide alternating current

Energy Storage

The purpose of energy storage, in grid-connected or microgrid applications, is to increase stability in our energy systems. There are certain times when we demand more energy than other times. For example, winter mornings, when buildings must be heated for the coming day, summer afternoons when the sun is strongest and the most cooling is needed, and the 5-7 p.m. period when office buildings and factories are still working and Americans are turning on the lights and appliances at home. All of these are times when energy demand rises. We call this instability "peak demand." When we look at renewable technologies like wind and solar, we have instability of supply. The wind blows sometimes and doesn't at other times. The sun shines sometimes and is hidden by clouds other times. It almost never shines at night.

Currently, we address demand instability with real-time pricing—making it more expensive to use power during peak demand periods. This is a pricing mechanism electric utilities use to create incentives for users to avoid dramatic peaks in demand. Supplying energy at those peak times is most expensive for utilities because they need to own and maintain generation equipment that may only be used for brief periods during the year. This equipment generates no income for the utility the rest of the time. Instability of demand is an expensive proposition for all parties.

We address instability of supply from renewables by simply hooking solar and wind power to the larger grid. The small amount of intermittent power added to the grid means that its effect is diluted and poses no real threat to system stability. As the portion of energy generated by intermittent renewables increases, however, there will be greater concern over its impact on the stability of the energy system. More to the point, a community that goes off the grid in island mode because of a grid outage will not have these avenues of mitigation.

This is where energy storage comes in. If we can store energy when demand is low or supply is high, and use that energy when demand increases or supply begins to drop off, we will reduce instability in the grid and increase the efficiency of our systems. From a distributed generation perspective, this means being able to supply more of our power with less on-site generation resources. The two most practical technologies to address these issues are batteries and thermal storage. There are other energy storage technologies such as fly wheels, pumped water and compressed air, but these technologies are not going to be as broadly applicable for community-level energy storage needs, so they are not addressed here.

Battery Storage

Energy storage with batteries is ideal for improving grid stability and energy resilience and empowering the widespread deployment of renewables. Batteries can improve power quality, allow for load shifting, and level the peaks and troughs of intermittent power sources; however, batteries still face a number of technical challenges. There are challenges associated with how much we can store and how quickly can it be discharged. There are also lifecycle considerations like maintenance and disposal. Industry and government labs are making impressive strides in all of these areas.

These challenges contribute to the biggest hurdle for building or grid scale energy storage: cost. Batteries are very expensive, costing between $1,700 and $4,900 per installed kilowatt as of 2010. On the bright side, many organizations are realizing strong paybacks by using batteries for a combination of peak shifting and demand response. As demands on an increasingly fragile grid increase and utilities continue to decommission older nuclear and coal plants, this type of demand-response programs is likely to increase, making battery storage cost effective in a number of applications.

How Battery Storage Works

All batteries rely on an electrochemical reaction between its two sides—the anode and the cathode. Oxidation on the anode side generates electrons that then pass through an electrolyte to the cathode side. A reduction reaction occurs on the cathode side where these electrons are absorbed. This process will continue until one of the electrodes runs out of a substance necessary to fuel the reaction. Charging of a battery is simply the reversal of this process.

There are two primary batteries for commercial use today on the building or campus scale: lead acid and lithium ion. Lead acid batteries are bigger and heavier. They are good for longer-term baseload power to operate your systems where you do not need a high degree of variability in power. Lithium ion batteries are a newer technology. These are much lighter batteries and can be scaled down in size (hence their use in cell phones). They can tolerate a wider temperature range and last longer than lead acid. These batteries are more expensive but can discharge their energy quickly to improve power quality, support utility grids, and play a more dynamic role in demand response.[97, 98]

Engineers all over the world are working on solutions for better batteries. There are a number of potential avenues from liquid metal batteries to advances in lithium ion that have improved the performance of energy storage systems. Isentropic storage, for example, is a new battery technology that involves heating a tank of rocks to store energy with a reversible heat pump to turn the gas in those hot rocks back into electricity.[99]

Where It Works Best

Batteries work best where they support other technologies (like wind and solar) or utility demand programs. The ability of batteries to turn intermittent loads into what are essentially base loads is a hugely valuable application. The ability of batteries to reduce a facility's demand as the grid approaches peak demand is valuable both for resilience and in garnering lucrative incentives from utilities.

Pros:
- Supports the use of intermittent power sources
- Can support demand-response programs and reduce your peak load
- Extends backup power time during a grid outage

Cons:
- These systems can be very expensive
- Batteries require space and electrical and physical infrastructure
- There is some efficiency loss in energy storage with batteries

Thermal Storage

Thermal storage is, simply put, the creation of ice (using your chiller system) during off-peak hours so you have access to free cooling during peak hours. This is not a technology that produces or stores electricity. However, the impact of thermal storage on energy resilience (particularly in warmer climates) is significant. Thermal storage allows clients to install smaller chiller systems in their buildings. Those smaller chillers, operating at 100 percent with supplemental chilling from the ice storage makes for a much more efficient system than having a large chiller running at partial load. Alternately, some facilities are using thermal storage for demand response—waiting for a curtailment request from a utility and then shutting down all chillers, drawing all cooling from their ice system. Whatever scheme is used, thermal energy storage can be very effective in reducing costs and increasing resilience in environments with a high level of cooling-degree-days.

How Thermal Storage Works

Thermal storage is a simple technology. It uses your chiller to cool a transfer medium (a water/chemical mixture or glycol) that then travels through a heat exchanger in a tank (or tanks) of water. The water is slowly turned to ice. When additional cooling is needed for the building, the transfer medium is run through a separate heat exchanger in the HVAC system where fans blow air over the cold pipes, creating cooled, conditioned air.

Where It Works Best

Thermal storage is essentially a load shifting technology, so the highest paybacks for these systems will be in areas with high demand charges and expensive real-time pricing. These systems work best in large buildings with high cooling loads—particularly in warm weather climates.

Pros:
- Can support demand-response programs and reduce your peak load
- Allows most users to reduce the size and increase the efficiency of their chiller systems
- Creates a useful load for on-site generation during off-peak hours
- Creates cooling with very little energy during a grid outage

Cons:
- Not as applicable in colder climates
- Water tanks require space and plumbing infrastructure
- There is some efficiency loss in the thermal storage process

MICROGRIDS

The previous section described various types of on-site generation technologies, their advantages and appropriate applications. For an individual building, one or more of these technologies can be installed and deployed fairly simply. When we are dealing with a multi-building community, the use of on-site generation becomes more complicated. You need to be able to balance the loads of several generation types, store energy for later use, and direct energy to critical systems when the external grid goes down. For this, you need what's known as a "microgrid."

Microgrids are small-scale versions of a centralized energy system. They are designed to operate in parallel with or inde-

ON-SITE GENERATION AND MICROGRIDS 171

pendent of the power grid, serving a building or community of buildings. The National Renewable Energy Laboratory describes microgrids as "power systems that are co-located with loads, regardless of the aggregated generation capacity and grid interconnection." This definition covers a large application space ranging from remote rural electrification and residential/community power networks to commercial, industrial, municipal, hospital, campus, and military base power grids. These applications also vary. Some are focused on cost of electricity (e.g. peak shaving), some are focused on local resource use (e.g. wind, solar, biomass systems), and some are focused on energy reliability and security (therefore, sophisticated generation and load controls are required)."[100] The microgrid is the stage upon which the energy generation and storage technologies discussed in the previous chapter can perform. Power control systems manage quality and capacity among the microgrid's microcontrollers, electronic interfaces, and process analytics, allowing administrators to manage the various components of a microgrid as a system. Microgrids provide your community with one integrated and reliable package to address your energy needs.

Without a microgrid, all on-site generation would need to be shut down in the case of a black out. This shutdown would be necessary to prevent uncontrolled electricity from entering distribution or transmission lines and creating an electrocution risk for line repair crews. The microgrid creates a single interface with the larger grid that can be severed when necessary. It not only allows users to control how energy is used within the microgrid, it can cleanly sever the connection with the larger grid so on-site generation is available for on-site sources and poses no danger to others. This island mode operation, as mentioned earlier in this book, allows the microgrid to continue generating and distributing energy without the larger grid. Whether that status means full, normal operation or powering a set of prioritized emergency loads, the microgrid gives users the control necessary to connect on-site generation and energy-storage sources with their critical loads.

How It Works

To understand microgrids, it helps to first think about the macrogrid. There are large power plants generating power all the time and smaller plants that start up when the demand becomes too great. There are long transmission lines and distribution networks to get the power to end users. There is a central control system that manages voltages and attempts to balance the load across the system. A microgrid has all of these components except the long-range transmission lines.

A microgrid embraces the entire suite of technologies at a site, allowing an end user to generate, store, control, and distribute power on-site. The components typically include:

- Power switches—These are the interconnections between the local microgrid and the power distribution from the utility grid. These systems are more than just a simple switch. They must match power frequency with the grid feed so there will not be power quality problems with disconnecting or reconnecting with the grid.

- Generation sources—On-site generation includes all of the generation technologies described in this chapter. Typically, on-site generation resources are scaled to power just the loads connected to the microgrid.

- Power storage—As discussed in the section on battery storage above, these systems are critical for improving power quality and leveling out the load provided by intermittent power sources.

- Rectifiers—Energy loads in the United States operate on alternating current. Direct current power from generation sources like solar panels needs to be rectified in order to provide alternating current to a microgrid.

- Power control system—Just as the utility has a grid control system, a microgrid needs a control system as well. This combination of hardware and software enables the monitoring of loads, generation sources, and storage. It coordinates the flow of power within the system. In advanced systems, this element can determine which base loads to connect or disconnect from the microgrid based on available power from generation sources. This infrastructure ensures that whatever power is produced is delivered to the most critical loads with sufficient power quality.

- Distribution lines—Connecting local generation sources to multiple buildings frequently requires the addition of power connections between those facilities. For communities that already connect to the grid in very discrete ways (like military bases or college campuses), this can be simple. For communities integrated into the power grid in a more decentralized way (like local governments), this connection is more complex and will likely require the addition of new power lines. New lines can be installed, but involve regulatory approval and engagement with the local utility.

The Value of Microgrids

The value of microgrids for energy resilience lies in their ability to keep on-site generation and storage technologies running during a grid outage, and to deliver power to the facilities that need it most. These are complex and expensive systems and the pursuit of energy resilience alone may not drive widespread investment in microgrids. However, microgrids can also create big savings on utility bills during normal operation. Microgrids allow clients to replace purchased electricity generated at inefficient coal plants and then transmitted over miles of cable with power generated at the point of usage. In places with high electricity rates and relatively low rates for other resources (because of low fuel prices or incentives for renewables), generating on-site energy can reduce total energy expenditures. See Figure 6-6.

Through the utilization of on-site generation or battery stor-

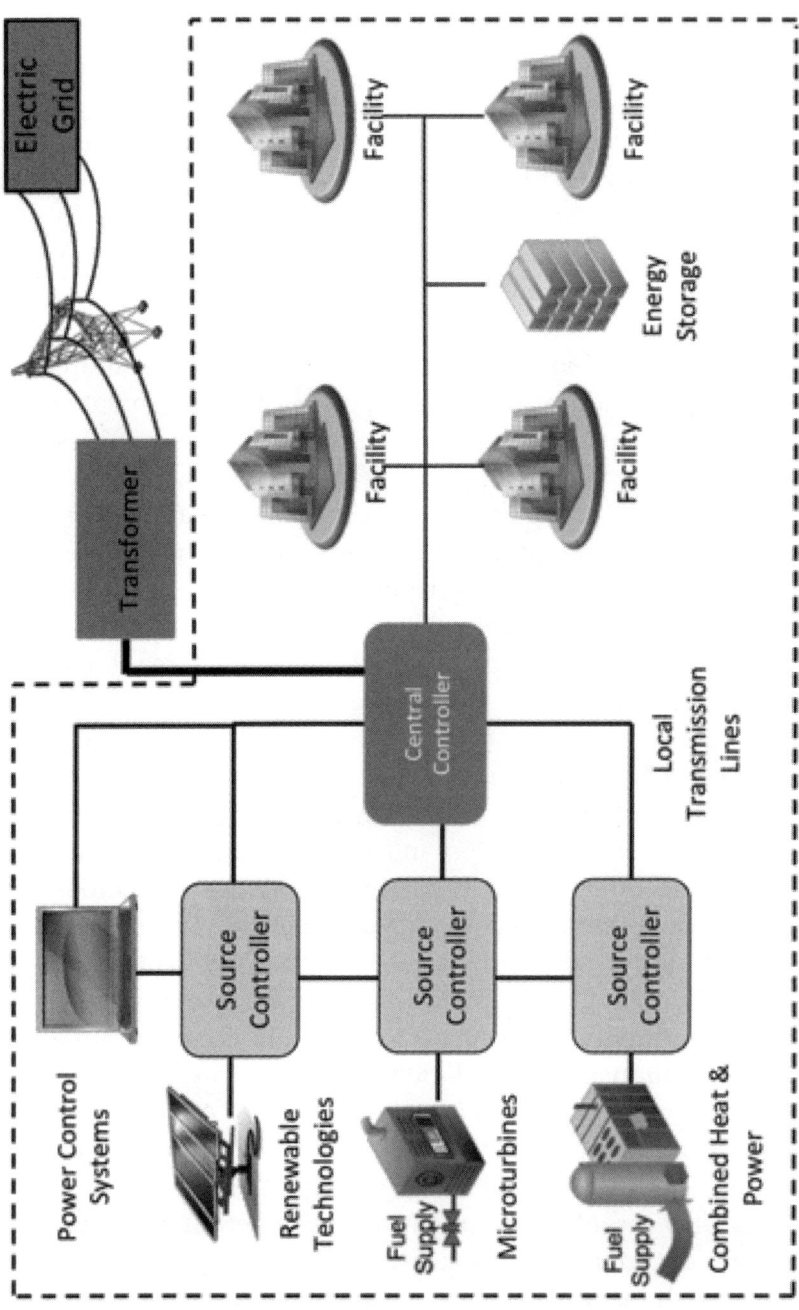

Figure 6-6: Example of a Microgrid System.

age that lowers demand on the utility lines during peak hours, communities can lower pricey demand charges. Most utilities provide considerable economic incentives to customers that can curtail load at certain times. Microgrids with power storage can also be used by the utility to reduce fluctuations in the local energy grid—generating additional payments from the utility for this service.

Microgrids eliminate many of the barriers to entry for on-site renewables such as solar and wind energy. Instead of being limited to using solar power on a given building or house, a microgrid allows a user to implement solar energy at the most ideal locations and use that energy anywhere within the microgrid, taking full advantage of every kilowatt of renewable energy generation available without needing to connect them to battery backup systems.

Microgrids (particularly those that incorporate battery storage or electric vehicle charging stations) also provide a facility or community with more flexibility in how and when they use energy from the grid. This means they can generate energy when it is cheapest and avoid using energy when it is most expensive. If, for instance, the cost of electricity is high and natural gas is low, on-site operation of a CHP plant will be optimal. If natural gas prices soar, the client can draw more from the grid.

While on-site generation may not currently be a cost-saving measure for every institution, the fact is that distributed-energy technologies are improving and dropping in cost, while the cost of electricity delivered from centralized plants is only expected to rise. Going forward, more and more institutions and communities will find that energy prices, incentives, and available technologies will justify the installation of a microgrid based on cost savings alone.

Where Microgrids Make Sense

Microgrids are ideally suited for energy management at the campus or community scale. A single building can typically incorporate on-site generation or backup power without the need for a microgrid. When multiple buildings are involved, the microgrid is necessary to employ these technologies for peak resilience and efficiency. Generally, these systems make sense where:

- Reliable, high quality energy is critical for mission success;
- Financial incentives make deploying renewables and other on-site generation or storage technologies attractive; and
- The regulatory and utility environment will allow for on-site generation and periodic disconnection from the grid during an outage event.

As discussed in more depth in Chapter 2, ideal settings for microgrids today include local governments, military bases, college campuses, and multi-building medical centers. As the challenges discussed in Chapter 1 (problems with grid reliability, extreme weather events, energy costs) increase, microgrids will make sense in a broader array of portfolios. This book focuses on domestic applications for microgrids. However, microgrids may also be an ideal solution for remote areas and developing nations—where deploying microgrids may be much more cost effective than building long-range transmission lines.

The case study below, focusing on New York University, provides an outstanding example of how on-site generation was used in a microgrid to support energy resilience in the wake of Hurricane Sandy. It is important to note, however, that these resources were only as valuable as they were to NYU due to thoughtful and integrated emergency planning.

Case Study: New York University[101]

New York University (NYU) is the nation's largest private university serving more than 38,000 students in 12 million square feet of building space (with plans to add another 2 million to that total). With a $50 million annual energy buy and a commitment to reducing its environmental impact, NYU has taken serious steps toward energy resilience. Their efforts include technical retrofits, like use of Telconet sensors to control lighting and HVAC—a measure that has saved an estimated 30 percent of energy costs in dormitories. They have also actively pursued management strategies including developing an entirely different HVAC and lighting schedule for the summer months when building uses change significantly. Finally, the school has engaged students with aggressive education programs around energy waste for incoming freshmen and an "NYU Unplugged" dorm competition.

Probably the biggest efficiency effort so far, though, has been the 2007-2010 installation of a new combined heat and power (CHP or 'cogen') central plant. In developing the new plant, the NYU team focused on practical needs, energy independence, sustainability, and reduction of energy consumption. The new plant reduced energy consumption at the service buildings by 25 percent and greenhouse gasses by 30 percent. There are backup feeds from ConEdison developed to ensure security if NYU's plant goes down. The frequency and power factor of NYU's system was designed to match ConEd's lines so the school can switch over with no interruption. At night, when the cogen plant makes more than needed, NYU gets a wholesale rate for energy it supplies back to the grid.

In October of 2012, when Hurricane Sandy hit the New York City area, lower Manhattan went dark for several days. The only island of light, heat and power was a collection of buildings on the campus of NYU. Their cogen plant continued to function as designed, and it kept students, faculty and most of the school's research projects comfortable. As John Bradley, Associate Vice President for Sustainability, Energy & Technical Services explains, "Security of the power system at NYU is critical because not only are the students and faculty relying on us, we have extremely important research that would be ruined if we can't maintain very specific conditions."

The cogen system plant serves some of the campus but not all. Additional connections are in place and can be made in the event of a grid shut-down so additional buildings can be fed from the single cogen plant. One example is the sports complex. Typically fed by the grid due to its large and fluctuating power needs, the

> sports complex was connected to the cogen plant after Sandy so it could be used as emergency shelter and to provide hot showers. Post-Sandy, NYU has plans to create more electrical connections and install more backup generation so the school will be that much more flexible and prepared in the event of another grid failure.
>
> NYU is an outstanding example of community energy resilience in action, chiefly because their systems were not developed solely to achieve independence, but to save money and increase performance for the school. With thoughtful planning, NYU leadership was able to achieve their cost-savings and sustainability goals while also giving themselves the one thing needed most in an emergency situation—power.

Clearly, there are multiple benefits to microgrids, including being prepared for the worst. The next chapter details the concepts of emergency planning for energy resilience.

Chapter 7

Planning for an Energy Emergency

Every institutional administrator must be prepared to respond to a loss of power from the regional grid. As we discussed in Chapter 1, threats to our energy system are real, and they are mounting. The professionals operating and maintaining our power delivery system have done an excellent job providing us with reliable electricity. Outages do happen, and with aging infrastructure, a growing threat of sabotage, and more frequent and severe storms, blackouts are bound to become more frequent. Chapters 5 and 6 described strategies that can lessen the impact of a blackout on your community through efficiency and on-site generation. This chapter will describe the process of planning for a power outage, prioritizing and protecting the core functions of your institution, and organizing human resources to manage an energy emergency.

As discussed in Chapter 2, the safety of constituents is central to the core missions at each of our four community types. In a blackout, hospitals must continue to treat patients and give staff the resources they need to do their jobs. A college must protect the safety of its residents, a military base must protect its soldiers while being ready to respond to any attack, and a town must ensure first responders are able to do their jobs and ensure the safety of the citizens. In order to perform these essential functions, an institution needs more than a lean energy profile and a robust on-site generation portfolio. Careful preparation is necessary to ensure, in the case of a grid outage or other energy-related event, backup power flows to the equipment that needs it most, key staff know how to activate a response plan, and responsibilities and lines of communication are clear, enabling staff to immediately begin managing operation and response.

Most institutions of the four types we examine in this book (educational campuses, healthcare campuses, military bases and municipalities) already have an emergency management plan in place. A written emergency action plan is required under Occupational Safety and Health Act regulations for all workplaces with more than 10 employees.[102] This chapter will not cover the process of emergency management in general, but instead focus on preparing for an energy emergency. We will use a general emergency planning framework to organize this discussion, but it is important for each community and institution to incorporate their power outage response plans into their existing emergency planning structure and policies. Although the guidance in this chapter applies to all four of our community types (and can be applied to other community types as well), we will primarily use examples from college campuses in an attempt to maintain some continuity in the hypothetical scenarios we raise.

RISK ASSESSMENT

An effective emergency management plan starts with a comprehensive risk assessment. This assessment consists of an evaluation of the threats and vulnerabilities at your institution. Each potential risk is evaluated based on both its likelihood and the potential effects on your facilities and core functions. This will give you a sense of the kinds of impacts you should try to prepare for, respond to, and recover from.

Step 1: Identify Threats
When examining threats to energy security at your institution, it may be useful to review the items described in Chapter 1, and consider the relevance of each to your institution. Remember, risk is a measure of severity and likelihood. If you have a micro-

EMERGENCY PLANNING   181

grid capable of going into island mode, it is possible a storm may affect your institution more severely than a grid outage caused by another factor, since it would have the potential to damage your own on-site generation and distribution equipment. You may decide, based on your geographical location, a strong storm is also more likely than a terrorist attack.

Create a list of threats to your energy supply, and rank them based on the severity of their impact, and the likelihood of that impact. This will allow you to identify the scenarios that pose the greatest energy risk at your institution. These are the scenarios you should focus on when creating an emergency plan for your energy system. Table 7-1 is a simplified example of a list of potential threats, and the emergency scenarios they might affect.

Step 2: Identify Vulnerabilities

Once potential threats are identified, the next step is to assess vulnerabilities at your institution. Each physical asset and function at your facility will be vulnerable to an energy shock in a different way. It's useful to break these vulnerabilities down into categories: facilities and grounds, core functions, and emergency functions.

Starting with facilities and grounds, review your asset list and consider how the different energy system threats you identified would affect each. On a college campus, some buildings may house vital research that is climate controlled, or otherwise dependent on a constant supply of energy. In some climates, the safety of student residents may depend on heating and cooling in residential buildings and cafeterias. Even outdoors, electric lights may play a vital role in ensuring the safety of students and staff as they walk the campus at night. An essential part of this inventory of facilities vulnerabilities will be those related directly to your energy infrastructure. Evaluate which parts of your on-site energy system are susceptible to the threats you identified, including on-site generation, distribution, electrical work within buildings. In any scenario where your on-site generation is less than your total campus demand, identify which parts of the system need to oper-

Table 7-1: Energy threat risks and impact scenarios (1 being the least severe/least likely):

Threat	Severity Ranking	Likelihood Ranking	Impact Scenario
Severe storm	5	2	Primary utility power out for 3-7 days; on-site energy infrastructure damaged
Moderate storm	2	5	Primary utility power out for less than 12 hours
Utility power out due to equipment failure/ overwhelming demand	3	4	Primary utility power out for less than 24 hours
Utility power out due to sabotage	4	1	Primary utility power out for 2-3 days; possible damage to on-site energy infrastructure
Drop in utility-provided power quality	2	4	Low power quality damages or shortens the operational life of major equipment
Energy Price Fluctuation	1	3	Disruption to operating budget due to energy price spike

ate while you are in island mode, and what will stop functioning if the regional grid goes down. If you have backup generators for individual facilities, note their location and how long they will remain operable in each location.

Next, examine the vulnerabilities specifically related to the functions of your institution. Start with the core mission. To continue with the case of a college campus, the core mission would be education and research. This core mission involves many functions, including classroom instruction, the maintenance of laboratories and research facilities, and the management of libraries and other educational resources. Of course, a college performs many other functions, including those related to housing, catering, healthcare, athletics, administration, information technology, and utilities (energy and water). For each of these functions consider the impact of the scenarios you identified in Step 1. Each function relies on a different set of facilities, equipment, and personnel. Perhaps a particular facility is not included in your facility's microgrid, leaving it in the dark even when the rest of the institution remains powered in island mode. How can you plan to continue to support the role that facility normally plays? Should all backup generators come on as a response to a power outage, or does it make sense to stagger their use in anticipation of a longer event? How do the roles and responsibilities of the staff change in this reduced energy scenario?

Finally, repeat this process for all the functions at the university directly related to emergency response. It is worth taking the time to separate these functions out, as understanding the vulnerabilities that affect them will inform how they are included in emergency response plans. It does no good to make a particular piece of equipment or staff person an essential piece of your institutions strategy to address an energy emergency if the availability of that resource will be compromised by the emergency itself. Emergency support functions may include fire services, medical services, emergency management, and utilities management. As with other functions, each of these relies on specific infrastructure and personnel. Each emergency support function should be analyzed against the scenarios identified in Step 1 to identify vulnerabilities and assess how that function might perform if compromised either by the same event that caused the energy shock, or the energy shock itself.

Table 7-2 depicts a highly simplified version of a risk assessment matrix for energy shocks at a college campus. The first column contains the impact scenarios from Step 1. The other columns list a few potential vulnerabilities in the categories of facilities, core functions, and emergency functions. For brevity, this table does not list every facility and function, but only gives a few examples for each scenario.

Note: For a real life example, you can review the document "Overview of System Failures—A Guidance for Incident Commander/Hospital Command Center," from the California Hospital Association, and available on their website. This guidance document lists utility loss scenarios (such as "normal power failure," "emergency power failure," and "steam failure) along with those functions in a hospital most likely to be affected by each, and instructions for how to respond to the emergency. (California Hospital Association, 2014).

In all stages of the risk assessment process, it is important to rely on existing sources of data. Often, institutions conduct general risk assessments as part of routine emergency planning. The structure of these evaluations should be adopted for energy risk assessment, and any information they contain about threats and vulnerabilities relevant to energy emergencies should be brought to bear. Other sources of relevant data include energy metering history for each facility (which you would have gathered in your efficiency efforts), nameplate and capacity factor information for on-site generation and distribution equipment, and information from existing emergency plans describing institutional emergency functions. Capturing a full list of these vulnerabilities will ensure a complete risk assessment and support proper preparation for an energy emergency.

Emergency Planning

Table 7-2: Energy Shock Impact Scenarios and Vulnerabilities

Impact Scenario	Facility Vulnerabilities	Core Function Vulnerabilities	Emergency Function Vulnerabilities
Primary utility power out for 3-7 days. On-site energy infrastructure damaged.	If microgrid generation or distribution is damaged, the campus will have to rely on small backup generators. Backup generation in west campus dorms will only last two days.	Students living on west campus are displaced. Cafeterias on west campus can no longer serve displaced students.	Damaged microgrid not able to power campus in island mode.
Primary utility power out for less than 12 hours.	Classrooms and libraries temporarily darkened.	Classes must be cancelled for part of a day.	Utility maintenance staff members have trouble getting to work due to storm damage.
Primary utility power out for less than 24 hours.	Air conditioning systems are off for a full day during the hottest time of the year.	Campus buildings become uncomfortably hot, and special cooling centers must be set up in buildings powered by backup generators.	The usable space in the student healthcare center is limited by the capacity of backup generators to cool it.
Primary utility power out for 2-3 days. Possible damage to on-site energy infrastructure	Microgrid on campus is damaged, and must be repaired before distribution equipment can bring power to campus assets.	Many core functions of the college are interrupted.	Emergency personnel are occupied securing the campus following sabotage event, and are not available to focus on energy emergency.
Drop in utility-provided power quality	Poor power quality damages equipment needed to operate critical facilities.	Loss of motors or HVAC systems disrupts campus life and classroom environments.	Loss of HVAC or pumps puts emergency response facilities at risk.
Disruption to operating budget due to energy price spike.	Facility vulnerabilities are low.	Constrained operating budget limits the amount of funding that can be dedicated to core functions.	Constrained operating budget limits the extra resources that can be brought on-line in case of a concurrent energy emergency.

THE FOUR PHASES OF EMERGENCY MANAGEMENT

Emergency management is typically described as consisting of four phases: mitigation, preparedness, response, and recovery. The mitigation phase is sometimes known as "prevention," but since there is little institutional administrators can do to prevent the kinds of energy shocks we consider in this book, we will instead focus on how to mitigate their effects. Although the cycle depicted in the figure below should recur after each emergency situation, the phases are highly interdependent, and you should revisit plans regularly, regardless of whether an emergency arises.[103]

Figure 7-1: The Four Phases of Emergency Management

Most institutions have already adopted this structure for their emergency planning activities. It is important to ensure preparations for energy emergencies are brought within this structure, and given careful attention in each phase. A power outage, for example, is often included as a potential contingency in an institutional emergency management plan. Of course, the energy system shocks we have discussed in this book are not limited to routine blackouts, and a truly energy resilient institution should include all relevant energy system shocks in its emergency management plan (including price fluctuations, regional outages, on-site equipment failures, etc.). The following section will give guidance on how to plan for a variety of energy contingencies in each phase of the four-phase cycle.

MITIGATION

All of the steps described in Chapters 5 and 6 fall into the category of mitigation, since the more an institution does to minimize its energy demand and optimize its on-site generation, the more it mitigates the effects of an energy shock. For example, if through building envelope improvements the R-value of a building is significantly improved, that structure will be able to withstand a longer period without power to its HVAC system before becoming too hot or too cold. If outdoor safety lighting is powered by integrated PV, the lights can stay on through the night even if the rest of the community goes dark.

On the generation side, a microgrid capable of going into island mode goes a very long way toward mitigating the effects of an energy emergency, since it allows on-site generation to continue to power an institution when the regional grid goes out. Even without island mode, a microgrid could help to mitigate the effects of a price spike, by allowing an institution to rely more heavily on one kind of fuel over another. If the price of coal power from the utility suddenly rises, energy managers could decide to rely more heavily on on-site natural gas turbines for power.

Another important tactic for mitigating the effects of an energy emergency is to ensure that institutional infrastructure, including building systems, energy systems, and emergency response systems (such as fire suppression), is operating properly and is in good repair. A program of continuous commissioning and regular inspections will ensure administration and maintenance staff have systems they can count on in an emergency.

PREPARATION

In the preparation phase, leaders define the policies and procedures to be followed in case of emergency. These policies will describe the central management structure that will respond to the emergency—often referred to as the Incident Command System (ICS). The Federal Emergency Management Agency lays out guidelines for the development of an ICS, and defines it as "a management system designed to enable effective and efficient domestic incident management by integrating a combination of facilities, equipment, personnel, procedures, and communications operating within a common organizational structure." An ICS is typically structured around five major functional areas: command, operations, planning, logistics, intelligence, and administration.

Energy emergency plans at your institutions should align with the practices for preparing for other types of emergencies, and response procedures should be integrated into your institution's existing ICS. While an energy emergency may occur on its own, it is more likely to arise as a part of a broader emergency event, such as a severe storm or a heat wave. In these cases, emergency response actions will need to be coordinated to ensure the resources are being prioritized and assigned appropriately to protect the vital functions of the institution.

Cornell University is an example of an institution that has taken emergency preparedness seriously and incorporated energy concerns into its planning efforts. The Cornell University Emergency Operations Plan defines emergency response priorities based on the following criteria:[105]

Priority 1: Protect and save the lives of University faculty, staff, students, visitors, and emergency responders
Priority 2: Preserve research including animals
Priority 3: Preserve University property and structures
Priority 4: Preserve the environment
Priority 5: Restore academic and business operations

Emergency Planning

Cornell also defines priorities for actions taken to protect university property:

Priority 1: Occupied buildings used by dependent populations that cannot be safely evacuated
Priority 2: Buildings critical to health and safety
Priority 3: Facilities that sustain the emergency response and recovery
Priority 4: Unoccupied research and classroom facilities and buildings
Priority 5: Unoccupied administrative buildings

These prioritization schemes are written to apply to any type of emergency, but they are highly applicable to the process of preparing for an energy emergency. Emergency resources, from staff to equipment and supplies, will be in finite supply during an emergency, and the ICS staff should have a clear plan for how to deploy those resources to address an energy emergency. This could mean sending utility maintenance staff to repair damaged electrical work in a building used to house emergency medical services before sending them to work on restoring power to a library. It may mean a limited number of diesel generators are allocated to protect live animal research before they are used for comfort cooling in a cafeteria. For more detail on this process, refer back to Step 4 in the section on "Evaluating Your Power Needs" in Chapter 6.

This kind of careful prioritization can also aid in preparations for an energy emergency by ensuring resources are allocated appropriately ahead of time. The installation of backup generators can be prioritized at buildings used to protect the safety of residents. When developing a microgrid, facilities that support high priority functions can be given priority for inclusion in a system capable of island mode. Highest priority facilities and systems may be given a more frequent commissioning schedule, or be subject to more frequent inspections.

Maintenance staff and emergency response staff should be

thoroughly and regularly trained on all aspects of the institution's emergency management plan, including those specifically related to the protection and restoration of the energy supply in an emergency. Staff should know what the chain of command is during an emergency, and what their specific responsibilities will be.

If back up generators are part of your preparation plan, know how long those generators will maintain your emergency-scenario load. This timing is critical to evaluate your community's resilience, but also for planning. This timing, and the steps to be taken when the generators run out of fuel, are critical components of the overall emergency management plan. Keep in mind that gasoline or diesel fuel that is stored for long periods may require special chemical treatment to keep it safe for use. Consult your fuel vendor for guidance on your particular situation. Emergency generators are likely to be seldom used, so ensure that maintenance staff is trained on the safe operation of your generator equipment. For more information on this, see the Energy Emergency Guidelines published by the Office of Electricity Delivery and Energy Reliability at the U.S. Department of Energy, available online at this address: http://energy.gov/oe/community-guidelines-energy-emergencies/using-backup-generators.

RESPONSE

The response phase occurs during an emergency. In this phase, all of the policies and procedures developed in the preparation phase go into action. Energy staff switch from their everyday responsibilities to their prescribed emergency situation roles. The ICS is convened, and begins to execute emergency procedures.

During the response phase, the energy management team will operate within the institution's emergency management structure, following established policies and prioritization to deliver power where it is needed most. It is important to note response to an energy emergency will not be the sole responsibility of energy professionals and emergency response staff. Staff in every facility will need to know what to do in their workspace if power, heat, or cooling goes out, and whom to notify to ensure coordination with the ICS. Many institutions provide workplace-specific power outage checklists along with other emergency action checklists to their staff, and then train those staff on the steps to be taken in case of a power outage. A quick internet search will turn up templates and examples of power outage checklists for each of the four community types. However, these checklists are most useful when they are tailored to the functions and equipment in each specific workplace.

RECOVERY

The recovery phase consists of efforts to restore the institution to normal operating status following an emergency. The steps involved in this phase depend on the specific qualities of the emergency. Generally speaking, the utilities or energy management division will be responsible for restoring the energy system, and ensuring all other functions at the university have the power they need to return to normal. Depending on the scope of the damage to the campus, administration may decide to convene a team to assess and catalog damage, so a plan can be made that includes prioritization of repair. If the energy emergency was caused by damage outside the institution, such as downed transmission lines, the recovery phase may include the maintenance of backup power on-site, and ensuring fuel supplies. Your recovery plan needs to include how and when you will reconnect to the grid and how you will transition back into full energy operations.

Good communication is critical in this phase. Institutional administrators must have good information from the energy management team on when to expect various resources to come back online, and which institutional functions can be supported at different stages of the recovery process. It is also important to keep constituents well informed of what facilities are affected by outages, where to access the services they need, and temporary conservation measures they must take. Remember that electronic communication and/or telephone service may be down so you may need alternate channels. Your energy plan needs to include how and when you will reconnect to the grid and how you will transition back into full energy operations.

The following case study describes an institution that has incorporated best practices from Chapters 5, 6 and 7 into its overall energy resilience strategy. Cornell University is aggressively mitigating the effects of an energy emergency through energy conservation measures, efficient operations, and a generous on-site generation portfolio. It has also engaged in a detailed, energy-specific emergency planning process.

Cornell University[106]

For more than a decade, energy management at Cornell has been central to a campus-wide commitment to sustainability. In 2009, the university codified its GHG reduction goals into an ambitious Cornell Climate Action Plan. This plan organized all of the university's climate change mitigation activities around five pillars: energy conservation, green building, renewable development, alternative transportation, and offsetting actions. Cornell exceeded the goals set in 2009 within two years, reducing its GHG emissions by 25 percent in the process, and decided to set new goals in an updated plan in 2011.

Despite these successes, when asked about what drives energy decisions at Cornell, sustainability is not the first thing Director of Energy Management Lanny Joyce mentions. "Reliability is our true bottom line," Joyce explains. "Environmental protection is a guiding principle for all we do, but at the end of the day, our top priority is to protect the business of the university." Evidence of this commitment not only to sustainability but to energy security can be seen just about anywhere you look at Cornell. Through an aggressive approach to energy conservation, on-site energy generation, and emergency management, Cornell has become a model of large-campus energy resilience.

Cornell's push toward an energy efficient and resilient campus began in earnest with the opening in 2000 of a $58.5 million lake-source cooling (LSC) project. The system consists of two water loops, one that circulates a closed loop of water from the university chilled water air conditioning system, and one that circulates water from the deep, cold water of nearby Cayuga Lake. The two loops meet at heat exchangers, and the two water loops never mix. "It would have cost us a lot less to simply replace every chiller on campus with brand new units," explains Joyce, "but the Lake Source Cooling project allowed us to take advantage of a natural resource, and reduced our peak loads tremendously. Right now our peak load is about 35 megawatts. Without Lake Source Cooling, our peak load would be 60 megawatts." The university reports that on average the LSC system saves more than 25 million kilowatt-hours of electricity per year to cool the Ithaca campus. This represents an energy savings of about 86 percent over conventional cooling technology.

Building on the success of LSC, Cornell has sought to exploit every opportunity to improve the energy efficiency of campus facilities, and reduce overall load. All new buildings at Cornell must be certified LEED silver or higher, and be 30-50 percent below the ASHRAE energy code in terms of energy intensity. The university has also sought to maximize the efficiency of existing buildings. Joyce explains, "Between 2001 and 2008 we spent $10 million on retrofitting existing buildings for energy efficiency. We will spend an additional $33 million by the end of 2015 on retrofit projects selected based on specific payback criteria. All of these investments are protected and maximized through a program of thorough continuous commissioning. Before 2001, Cornell's buildings were not continuously commissioned. Now we have 10 full time employees working on all 14 million square feet of conditioned space at all times."

To meet the reduced energy demand achieved through these efforts, the university relies on a combination of on-site generation technologies. The centerpiece of this system is the Cornell Central Energy Plant, a 38 megawatt combined heat and power plant. Cornell also owns and operates a 1 megawatt hydro plant that dates back to the founding of the University in 1865. As part of a successful effort to completely remove coal from Cornell's energy portfolio by 2011, the university retrofitted the plant in 2009 with two 15 megawatt natural gas turbines. Before the installation of the new turbines, the university purchased 85 percent of its electricity from the local power utility. Now, over 80 percent of Cornell's annual power needs are met with the CHP plant. The steam generated in the new turbines heats 150 campus buildings. The campus also has several solar installations and a small 1 megawatt hydropower plant in one of its famous gorges. "In addition to getting our heat for 'free' from the CHP plant," explains Joyce, "we are producing up to 39 megawatts of energy right on campus. At our winter minimum electric load of about 20 megawatts during intersession, we can export up to 19 megawatts of power back to the utility." Importantly for energy resilience, the 2009 retrofit of the CHP plant allowed Cornell to go into island mode when the regional grid is down, relying on on-site systems for the vast majority of its power, heating, and cooling needs. In 2014, Cornell brought on line its first third party owned utility scale 2 megawatt solar electric remote net metered photovoltaic facility that will provide about 1% of annual campus electricity needs.

Cornell's achievements in energy efficiency and on-site generation represent only two of the three pillars of energy resilience, but Cornell has achieved in the third category as well, establishing a comprehensive emergency management system. The University maintains separate written plans for emergency and mitigation, emergency preparedness, emergency operations, and recovery. In the detailed emergency operating plan, the university designates 18 emergency support functions, including fire services, housing, dining, health services, utilities, and of course emergency management. The plan describes the responsibilities for staff in each of these functions in terms of emergency preparedness, response, and recovery. The plan describes potential vulnerabilities to the universities on-site energy resources and gives guidance for prioritizing the universities functions when allocating limited energy resources.[105]

The efficiency of Cornell's infrastructure, the redundancy in its power systems, and the careful emergency management adopted by the university have together left the university's core mission well protected, and the institution highly resilient to energy shocks.

The Office of Electricity Delivery and Energy Reliability at the U.S. Department of Energy maintains Community Guidelines for Energy Emergencies, including advice tailored for businesses, homeowners, and local governments. These guidelines are a valuable resource for local communities and institutions that are beginning the process of preparing for an energy emergency, and are available online at this address: http://energy.gov/oe/services/energy-assurance/emergency-preparedness/community-guidelines-energy-emergencies.

The process of pursuing energy resilience is truly a process of emergency preparedness. Every action taken to improve an institution's facility efficiency, create a robust energy management team, and expand its on-site generation portfolio increases its ability to be resilient to the next energy shock. The management processes described in this chapter are designed to direct investment in energy resilience toward identified risk, and ensure selected projects will produce the highest possible return in terms of containing and resolving the energy emergencies most likely to arise. Remember to consult the Energy Resilience Maturity Model in Chapter 4 for guidance on benchmarking your performance in the area of emergency management, and the other areas of energy resilience management discussed in this book.

Conclusion

Energy independence is not a binary state, but a continuum. And each community, weighing the risks and action steps required, must decide where it will strive to operate along that continuum. Greater reliability on outside energy sources means more risk. Striving to reduce reliability on outside energy sources requires vision, long-term planning, and investment. The good news is that, for many communities, it is possible to move along that continuum far beyond today's status quo.

All of the critical elements of energy resilience outlined in this book have been piloted, evaluated, and found practical in multiple environments. Organizations can dramatically reduce their energy consumption using off-the-shelf, cost-effective technologies, smarter usage patterns, and best management practices. There is tremendous potential for on-site generation and microgrids to supply power during grid outage events and reduce costs even during normal operation. Taking smaller steps can also have beneficial effect and put your organization on the path toward being safer and more secure. The book has outlined how effective planning for emergencies allows us to make the most of our resources on the ground and limit the negative impacts of shocks to the energy system.

Over the course of this book we have provided details about specific strategies and technologies that can be used to change your institution's position on the energy resilience spectrum. Our goal has not been to assert a one-size-fits-all approach, but to provide specific steps and questions the reader can use to develop their own path toward energy resilience.

We have used case studies of other institutions to demonstrate how resilience strategies and technologies are being implemented in the real world. This final case study focuses on the Fort Hood military base in Killeen, TX. We selected Fort Hood as the final case study for this book because the leadership team there is clearly focusing on each of the three main pillars of energy resilience in their approach—efficiency, on-site generation, and emergency planning.

> **Fort Hood Case Study[107]**
>
> One out of every ten soldiers in the Army serves at Fort Hood. It is the largest active duty armored post in the United States, and the only post in the USA capable of supporting two full Army armored divisions. Supporting an operation that large requires an incredible amount of energy. In order to combat costs and maintain operational readiness, Fort Hood's leadership has implemented a holistic energy resilience program. Their program includes all three aspects of energy resilience focused on in this book—energy management, on-site generation, and emergency planning. Of every community evaluated for this book, Fort Hood seems to have the most comprehensive energy resilience program.
>
> The Fort Hood program began with a drive to reduce energy costs. They have leveraged energy savings performance contracts to replace outdated or inefficient technologies. They have coupled these technology retrofits with behavior modification. Brian Dosa, Fort Hood's Director of Public Works, explained, "You can put in super high efficiency technology, but if the people who use energy in the space aren't cognizant of your goals, you won't reach them." To that end, Fort Hood has begun submetering buildings in order to generate mock energy bills for each unit—helping commanders to understand energy usage and think more clearly about being stewards of their own resources. The team also provides energy conservation training to units that request it and has deployed home energy management system studies where on-base residents were offered a $50 gift card to connect their thermostat to a central data collection and control system.
>
> Fort Hood is also aggressively evaluating the deployment of more sustainable on-site generation. There is currently a 3,000 panel, one megawatt solar array providing electricity for an on-base residential village. Fort Hood's Public Works team is working with the Army's Energy Initiative Task Force to evaluate large solar and/or wind installations for the base. The base has limited

CONCLUSION 199

> emergency back-up power for key buildings currently and sees increased renewable energy as critical to the energy security of their operation.
>
> Beyond just checking these two boxes, Fort Hood is integrating their efficiency and generation assets to complement each other. Again, Brian Dosa: "We are coupling renewable technologies (like photovoltaic and solar thermal) with efficiency driven by our Utility Management and Control System. This system lets us reduce overall demand but also accomplish peak shaving." Fort Hood is also a pilot for a DOD electric vehicle program. This program includes acquiring 55 plug-in electric vehicles to test on the base. The batteries in the vehicles will be used to store electricity in low-demand times and then tapped as an alternate power source during peak demand times.
>
> While discussing emergency preparedness planning is not something military personnel are likely to do with an author, Fort Hood has made a serious, top-level effort to develop plans and education around their response to a loss of grid power. Fort Hood staff members see themselves as leaders—in every respect, including in the energy management field. Supported by the Energy Chief of the Army and agency-wide programs, Fort Hood has made dramatic strides in energy resilience. While not every community will have the same level of internal competence and external support, Fort Hood is an outstanding example of a holistic and comprehensive approach to energy resilience planning.

As the case studies in this book have demonstrated, there are a variety of ways your organization can move along the energy resilience spectrum. This book's final piece of advice is to attempt to implement your strategies and technologies in ways that leverage each other. Create results that are beyond the scope of any single policy or technology by coordinating these efforts in such a way that they support and reinforce each other. Reduce energy usage in ways that will make it easier to deploy on-site generation. Deploy generation assets so that they support your institution's critical needs during an emergency. Publicize your efforts so that success will beget more support and success. Use the resilience of your institution as a selling point for potential residents, employees, business partners and clients. Use energy resilience not as an insurance policy, but as a pathway to achieve your core goals.

Looking forward, it seems that the threats to our energy systems will continue to increase. Each new test of our nation's infrastructure or increase in geopolitical unrest will create new acolytes for energy resilience and bolster your case for organization-wide investments. Energy resilience will cease to be something talked about only at energy engineering conferences or by urban planning wonks. It will be another common metric by which we value safety, desirability, and economic competitiveness. We, the authors of this book, believe that conventional utility business models will rapidly shift, and those energy companies that cannot adapt (delivering more resilient, responsive energy systems) will fail to be competitive. As technology continues to develop, distributed generation will become more than just cost competitive—it will be the most logical choice for most communities. Efficient, net-zero buildings will become commonplace and institutional (and even personal) responsibility for access to energy will become the norm. We, as a society, will move along the resilience curve.

It all begins with a conversation. One concerned team member talks to another about the risks to their institution from energy system shocks. They ask some "What if?" questions and land on the most important question, "What are we doing about this?" From there, others are brought into the conversation and it is raised to the leadership level. Using information from this book and found regularly in the news media, the team understands the threat and starts to understand how it might impact their institution. This team doesn't need to be frustrated by a lack of understanding around what to do next. You can combine the approaches outlined in this book with your own experience and expertise to lead your community to a better place along the energy resilience spectrum. We hope that you will do exactly that. We hope that you will start the conversation.

Citations

[1] Alyson Kenward, and Urooj Raja, "Blackout: Extreme Weather, Climate Change, and Power Outages," *Climate Central*, 2014. http://assets.climatecentral.org/pdfs/PowerOutages.pdf.

[2] "Do we have enough oil worldwide to meet our future needs?" *U.S. Energy Information Administration*, November 7, 2013. http://www.eia.gov/tools/faqs/faq.cfm?id=38&t=6

[3] "Annual Energy Outlook 2014: Early Release." *U.S. Energy Information Administration*, April 2013. http://www.eia.gov/forecasts/aeo/er/pdf/0383er(2014).pdf.

[4] "Annual Energy Outlook 2013, with Projections to 2040," *U.S. Energy Information Administration*, April, 2013. http://www.eia.gov/forecasts/archive/aeo13/pdf/0383(2013).pdf

[5] John Rogers, et al. "Water-smart power: Strengthening the U.S. electricity system in a warming world." Cambridge, MA: Union of Concerned Scientists, July, 2013.

[6] "Energy Demands on Water Resources: Report to Congress on the Interdependency of Energy and Water," US Department of Energy, December, 2006. http://www.sandia.gov/energy-water/docs/121-RptToCongress-EWwEIAcomments-FINAL.pdf]

[7] Thomas Brown, et al., "Projected freshwater withdrawals in the United States under a changing climate," *Water Resources Research*, Vol. 49, 1259-1276, 2013.

[8] J.F. Kenny, et al., "Estimated use of water in the United States in 2005," *U.S. Geological Survey Circular*, 1344, October 2009. http://pubs.usgs.gov/circ/1344/.

[9] Jim Witkin, "In a Hot, Thirsty Energy Business, Water is Prized," *The New York Times*, October 8, 2013. http://www.nytimes.com/2013/10/09/business/energy-environment/in-a-hot-thirsty-energy-business-water-is-prized.html?_r=0.

[10] "Effects of Climate Change on Energy Production and Use in the United States," *U.S. Climate Change Science Program and the subcommittee on Global change Research*, February 2008. http://downloads.globalchange.gov/sap/sap4-5/sap4-5-final-all.pdf.

[11] "Drought Impact on Hydropower Frequently Asked Questions," *California Energy Commission*. http://www.energy.ca.gov/drought/drought_FAQs.html.

[12] "Energy-Water Nexus: Coordinated Federal Approach Needed to Better Manage Energy and Water Tradeoffs," *U.S. Government Accountability Office*, 2012. http://www.gao.gov/products/GAO-12-880.

[13] "2013 Report Card for America's Infrastructure—Energy," *American Society of Civil Engineers*, 2013. http://www.infrastructurereportcard.org/.

[14] "Report Card for America's Infrastructure, 2009 Grades," *American Society of Civil Engineers*, 2009. http://www.infrastructurereportcard.org/.

[15] "Failure to Act—The Impact of Current Infrastructure Investment on America's Economic Future," *American Society of Civil Engineers*, 2012. http://www.asce.org/uploadedFiles/Infrastructure/Failure_to_Act/Failure_to_Act_Report.pdf.

[16] Massoud Amin, "Turning the Tide on Outages—What are the true costs of implementing—or failing to implement—a stronger, smarter and more robust grid," DRAFT, July 18, 2011. http://massoud-amin.umn.edu/publications/Turning_the_Tide_on_Outages_MA_Draft_07-18-2011.pdf.

[17] Massoud Amin, "Smart Grid—Safe, Secure, Self-Healing," *IEEE Power & Energy Magazine*, January/February 2012. http://magazine.ieee-pes.org/january-february-2012/smart-grid-safe-secure-self-healing/.

[18] Massoud Amin, "North American Electricity Infrastructure: System Security, Quality, Reliability, Availability, and Efficiency Challenges and their Societal Impacts," Chapter two in "Continuing Crises in National Transmission Infrastructure: Impacts and Options for Modernization," *National Science Foundation*. June 2004. http://massoud-amin.umn.edu/publications/NSF_Chapter2.pdf.]

[19] "Annual Energy Outlook 2014: Market Trends—Natural Gas," *U.S. Energy*

CITATIONS

Information Administration, April 7, 2014. http://www.eia.gov/forecasts/aeo/MT_naturalgas.cfm.

[20] "Pipeline Security," *U.S. Transportation Security Administration,* Updated November 13, 2013. https://www.tsa.gov/stakeholders/pipeline-security.

[21] Tim Boersma and Charles K. Ebinger, "Prevailing Debates Related to Natural Gas Infrastructure—Investments and Emissions," *The Brookings Institute,* January 2014. http://www.brookings.edu/~/media/research/files/reports/2014/natural%20gas%20infrastructure%20investments%20emissions/debates%20natural%20gas%20infrastructure%20investments%20emissions%20boersma%20ebinger.pdf.

[22] "Pipeline Security," *U.S. Transportation Security Administration,* Updated November 13, 2013. https://www.tsa.gov/stakeholders/pipeline-security.

[23] Jeff Tollefson, "Methane leaks erode green credentials of natural gas," *Nature,* January 2, 2013. http://www.nature.com/news/methane-leaks-erode-green-credentials-of-natural-gas-1.12123.

[24] Michael Wines, "Gas Leaks in Fracking Disputed in Study," *The New York Times,* September 16, 2013. http://www.nytimes.com/2013/09/17/us/gas-leaks-in-fracking-less-than-estimated.html?_r=0.

[25] "Report to Congressional Requesters—Climate Change—Energy Infrastructure Risks and Adaptation Efforts," *U.S. Government Accountability Office* GAO-14-74, January 2014. http://www.gao.gov/assets/670/660558.pdf.

[26] "U.S. Energy Sector Vulnerabilities to Climate Change and Extreme Weather," *U.S. Department of Energy,* July 2013. http://energy.gov/sites/prod/files/2013/07/f2/20130716-Energy%20Sector%20Vulnerabilities%20Report.pdf

[27] Kenneth Kunkel, et al., "Regional Climate Trends and Scenarios for the U.S. National Climate Assessment. Part 9. Climate of the United States." *National Oceanographic and Atmospheric Administration,* January, 2013. http://scenarios.globalchange.gov/sites/default/files/NOAA_NESDIS_Tech_Report_142-9-Climate_of_the_Contiguous_United_States_3.pdf.

[28] Andrea Thompson, "2014 on track to become the hottest year on record,"

Scientific American, September 23, 2014. http://www.scientificamerican.com/article/2014-on-track-to-be-hottest-year-on-record/.

[29] "Report to Congressional Requesters—Climate Change—Energy Infrastructure Risks and Adaptation Efforts," *U.S. Government Accountability Office*, January 2014. http://www.gao.gov/assets/670/660558.pdf.

[30] "U.S. Energy Sector Vulnerabilities to Climate Change and Extreme Weather," *U.S. Department of Energy*, July 2013. http://energy.gov/sites/prod/files/2013/07/f2/20130716-Energy%20Sector%20Vulnerabilities%20Report.pdf.

[31] "Climate Impacts on Water Resources," *U.S. Environmental Protection Agency*, Updated: August 13, 2014. http://www.epa.gov/climatechange/impacts-adaptation/water.html

[32] Ker Than, "Stanford Scientists Investigate Worst Drought in California's History," *Stanford News*, February 27, 2014. http://news.stanford.edu/news/2014/february/drought-climate-change-022714.html.

[33] "How Hurricane Sandy Affected the Fuels Industry," *NACS Online*, 2013. http://www.nacsonline.com/YourBusiness/FuelsReports/GasPrices_2013/Pages/How-Hurricane-Sandy-Affected-the-Fuels-Industry.aspx.

[34] Amanda Staudt and Renee Curry. "More Extreme Weather and the U.S. Energy Infrastructure." National Wildlife Federation. 2011. http://www.nwf.org/pdf/Extreme-Weather/Final_NWF_EnergyInfrastructureReport_4-8-11.pdf.

[35] "Climate Change Indicators in the United States," *U.S. Environmental Protection Agency*, Updated, May 2014. http://www.epa.gov/climatechange/science/indicators/oceans/sea-level.html.

[36] "Terrorism and the Electrical Power Distribution System," *National Research Council, National Academy of Sciences*, 2012. https://portal.mmowgli.nps.edu/c/document_library/get_file?uuid=f0f028d6-de39-45d1-8afc-2fb7d0204420&groupId=10156.

[37] Nicole Perlroth, "Russian Hackers Targeting Oil and Gas Companies," *The New York Times*, June 30, 2014. http://www.nytimes.com/2014/07/01/technology/energy-sector-faces-attacks-from-hackers-in-russia.html.

[38] George Markhoff, "Andy Grove Offers an Energy Solution," *The New York Times*, June 27, 2008, http://bits.blogs.nytimes.com/2008/06/27/andy-grove-offers-an-energy-solution/.

[39] Andy Grove and Robert Burgleman, "An Electric Plan for Energy Resilience," *The McKinsey Quarterly: Energy, Resources, Materials*, December, 2008.

[40] John M. Broder, "Climate Change Will Cause More Energy Breakdowns, US Warns." *The New York Times*, July 11, 2013. http://www.nytimes.com/2013/07/11/us/climate-change-will-cause-more-energy-breakdowns-us-warns.html.

[41] Andrew Revkin, "Dot Earth—How Natural Gas Kept Some Spots Bright and Warm as Sandy Blasted New York City," *The New York Times*, November 5, 2012. http://dotearth.blogs.nytimes.com/2012/11/05/how-natural-gas-kept-some-spots-bright-and-warm-as-sandy-blasted-new-york.

[42] David Goodman and Colin Moynihan, "Patients Evacuated From City Medical Center After Power Failure," *The New York Times*, October 30, 2012. http://www.nytimes.com/2012/10/30/nyregion/patients-evacuated-from-nyu-langone-after-power-failure.html.

[43] Andrew Zolli, "Op-ed: Learning to Bounce Back," *The New York Times*, November 2, 2012. http://www.nytimes.com/2012/11/03/opinion/forget-sustainability-its-about-resilience.html?pagewanted=all&_r=0.

[44] Jeff Rich, Executive Director, Gunderson Envision Program, Gundersen Health System, Interview with author, conducted May 21, 2013.

[45] Barry Rabner, President and CEO, Princeton HealthCare System, Interview with author, conducted May 21, 2013.

[46] Fred Beck and Eric Martinot, "Renewable Energy Policies and Barriers," in *Encyclopedia of Energy*, edited by Cleveland Culter (San Diego: Academic Press/Elsevier Science, 2004), 365-383.

[47] Ann Rappaport and Sarah Creighton, *Degrees That Matter: Climate Change and the University* (Boston: MIT Press, 2007).

[48] Sam Booth, et. al., "Net Zero Energy Military Installations: A Guide to Assessment and Planning," National Renewable Energy Laboratory, August, 2010. http://www.nrel.gov/docs/fy10osti/48876.pdf.

[49] S.B. Van Broekhoven, et. al. "Microgrid Study: Energy Security for DoD Installations, Technical Report of Lincoln Laboratory," Massachusetts Institute of Technology, June 18, 2012. http://www.dtic.mil/cgi-bin/GetTRDoc?AD=ADA565751.

[50] Roland Risser, "Energy Department's Hospital Energy Alliance Helps Partner Save Energy and Money," U.S. Department of Energy, September 4, 2012. http://www.energy.gov/articles/energy-department-s-hospital-energy-alliance-helps-partner-save-energy-and-money.

[51] "Department of Defense Annual Energy Management Report Fiscal Year 2011," *U.S. Department of Defense*, September, 2012. http://www.acq.osd.mil/ie/energy/library/FY.2011.AEMR.pdf.

[52] Kate Galbraith, "Despite Budget Cuts and Base Expansion, Trying to Save Energy." *The New York Times*, April 26, 2012. http://www.nytimes.com/2012/04/27/us/at-fort-bliss-and-fort-hood-going-solar-for-net-zero-energy-production.html?pagewanted=all&_r=0.

[53] Mark Jaffe, "Army Enlisting Help to Make its Bases Energy Independent," *The Denver Post*, June 3, 2012. http://www.denverpost.com/ci_20765449/army-enlisting-help-make-its-bases-energy-independent.

[54] Michael Heiman, "What Caused Generators to Fail at NYC Hospitals?" *CBS News*, October 28, 2012. http://www.cbsnews.com/8301-204_162-57544376/what-caused-generators-to-fail-at-nyc-hospitals/.

[55] Anemona Hartocollis, "A Flooded Mess That Was a Medical Gem." *The New York Times*, November 9, 2012. http://www.nytimes.com/2012/11/10/nyregion/damage-from-hurricane-sandy-could-cost-nyu-langone-millions.html.

[56] Alexandra Sifferlin, "Storm Destroys Valuable Medical Research," *Time Magazine*, November 2, 2012. http://healthland.time.com/2012/11/02/storm-destroys-valuable-medical-research/.

CITATIONS   207

[57] Matthew L. Wald, "How N.Y.U. Stayed (Partly) Warm and Lighted," *The New York Times, Green Blog*, November 5, 2012. http://green.blogs.nytimes.com/2012/11/05/how-n-y-u-stayed-partly-warm-and-lighted/.

[58] Lanny Joyce, Director of Energy Management, Cornell University, Interview with author, conducted July 30, 2013.

[59] D.W. Nutt, "Cornell Solar Power Project Gets Boost from State," *The Ithaca Journal*, July 13, 2013. http://www.sustainablecampus.cornell.edu/blogs/news/posts/cornell-solar-power-project-gets-boost-from-state.

[60] New York University, "NYU Switches on Green CoGen Plant and Powers Up for Sustainable Future [Press Release]," January 21, 2011. http://www.nyu.edu/about/news-publications/news/2011/01/21/nyu-switches-on-green-cogen-plant-and-powers-up-for-the-sustainable-future.html.

[61] Office of the Governor, State of New Jersey, "Christie Administration Announces New Hazard Mitigation Initiative to Support Post-Sandy Energy Resilience [Press release]," July 10, 2014. http://www.state.nj.us/governor/news/news/552014/approved/20140710b.html.

[62] Frank Morris, "Missouri Town Is Running On Vapor—And Thriving," National Public Radio, August 9, 2008. http://www.npr.org/templates/story/story.php?storyId=93208355.

[63] University of Missouri Extension, "Rock Port, Missouri, First 100 Percent Wind-powered Community In U.S." *ScienceDaily*, July 16 2008. www.sciencedaily.com/releases/2008/07/080715165441.htm.

[64] Christopher Russell, *The Industrial Energy Harvest*. Energy Pathfinder, LLC. 2008.

[65] "Energy and Cost Savings Calculators for Energy-Efficient Products," U.S. Department of Energy, Office of Energy Efficiency and Renewable Energy web site. http://www1.eere.energy.gov/femp/technologies/eep_eccalculators.html.

[66] "Cash Flow Opportunity Calculator," *U.S. Environmental Protection Agency ENERGY STAR Program*. http://www.energystar.gov/buildings/tools-and-resources/cash-flow-opportunity-calculator-excel.

[67] BJ Tomlinson, Energy Branch Chief, Fort Bliss, Interview with author, conducted July 1, 2013.

[68] "Revolving Credit—All Grown Up," *University of North Carolina Environmental Finance Center*, http://efc.web.unc.edu/2013/02/20/revolving-credit-all-grown-up/.

[69] Dominic Frongillo, Councilor and Deputy Town Supervisor, Caroline, New York, Interview with author, conducted May 16, 2013.

[70] "Energy Independence and Security Act of 2007: Section 432," *U.S. Government Printing Office*, pages 116-120. http://www.gpo.gov/fdsys/pkg/BILLS-110hr6enr/pdf/BILLS-110hr6enr.pdf.

[71] "Procedures for Commercial Building Energy Audits, 2nd Edition," ASHRAE, 2011. https://www.ashrae.org/resources—publications/bookstore/procedures-for-commercial-building-energy-audits.

[72] Evan Mills, "Building Commissioning: A Golden Opportunity for Reducing Energy Costs and Greenhouse Gas Emissions," *Lawrence Berkeley National Laboratory*, July 2009. http://cx.lbl.gov/documents/2009-assessment/lbnl-cx-cost-benefit.pdf.

[73] Donald Wulfinhoff, "Where Energy Efficiency and Alternative Energy Work, Where They Don't, and Why," *Strategic Planning for Energy and the Environment*, Vol 31, No. 2, Fall 2011.

[74] Hal Alguire, Director of Public Works, Fort Carson, Interview with author, June 17, 2013.

[75] Annual Energy Outlook 2014, *U.S. Energy Information Administration*, April 2014. http://www.eia.gov/forecasts/aeo/.

[76] W.R.L. Anderegg, "Expert Credibility in Climate Change," *Proceedings of the National Academy of Sciences*, Vol. 107 No. 27, 12107-12109, 21 June 2010. http://www.pnas.org/content/107/27/12107.abstract.

[77] P.T. Doran and M.K. Zimmerman, "Examining the Scientific Consensus on Climate Change," *Eos Transactions American Geophysical Union*, Vol. 90 Issue 3

(2009), 22. http://tigger.uic.edu/~pdoran/012009_Doran_final.pdf.

[78] "Toxic Air: The Case for Cleaning Up Coal-Fired Power Plants," *The American Lung Association*, March, 2011. http://www.lung.org/assets/documents/healthy-air/toxic-air-report.pdf.

[79] "The Potential Benefits of Distributed Generation and Rate-Related Issues That May Impede Their Expansion," *U.S. Department of Energy*, February 2007. http://energy.gov/sites/prod/files/oeprod/DocumentsandMedia/1817_Report_-final.pdf.

[80] "Commercial Building Energy Consumption Survey," *Energy Information Administration*, 2003 data. http://www.eia.gov/consumption/commercial/.

[81] Nancy Carlisle, Otto Van Geet, Shanti Pless, "Definition of a Zero Net Energy Community," *National Renewable Energy Laboratory*, November 2009. http://www.nrel.gov/docs/fy10osti/46065.pdf.

[82] "How Coal Gassification Power Plants Work." Department of Energy Office of Fossil Energy. http://energy.gov/fe/how-coal-gasification-power-plants-work.

[83] Noriko Behling, "Making Fuel Cells Work," *Issues in Science and Technology*, Issue 29-3, 12, 2013. http://www.fuelcells.org/pdfs/Behling.pdf.

[84] Brian C. H. Steele and Angelika Heinzel, "Materials for Fuel Cell Technologies," *Nature* 414, 345-352, November 15, 2001. http://www.nature.com/nature/journal/v414/n6861/full/414345a0.html.

[85] Andy Skok, "Fuel Cell CHP for Industrial & Critical Infrastructure Support," *Presentation at the World Energy Engineering Congress*, October 2, 2014.

[86] "State of the States: Fuel Cells in America 2013," *U.S. Department of Energy—Office of Energy Efficiency and Renewable Energy—Fuel Cell Technologies Office*, October 2013. http://www1.eere.energy.gov/hydrogenandfuelcells/pdfs/state_of_the_states_2013.pdf.

[87] "Combined Heat and Power: A Clean Energy Solution" *U.S. Department of Energy and U.S. Environmental Protection Agency*, August 2012. http://energy.gov/sites/prod/files/2013/11/f4/chp_clean_energy_solution.pdf.

[88] "Combined Heat and Power: Evaluating the Benefits of Greater Global Investment," *International Energy Agency*, March 2008. http://www.iea.org/publications/freepublications/publication/chp_report.pdf.

[89] U.S. EPA Web Site; Combined Heat and Power Partnership. www.epa.gov/chp/basic/efficiency.html.

[90] "Tracking the Sun VI: An Historical Summary of the Installed Price of Photovoltaics in the United States from 1998 to 2012," *Lawrence Berkley National Laboratory*, July 2013. http://emp.lbl.gov/sites/all/files/lbnl-6350e.pdf.

[91] Jordan Schneider and Rob Sargent "Lighting the Way," *Environment America Research & Policy Center*, August 2014. http://environmentamericacenter.org/sites/environment/files/reports/EA_Lightingtheway_scrn.pdf.

[92] "2012 Wind Technologies Market Report," *U.S. Department of Energy Office of Energy Efficiency and Renewable Energy*, August 2013. http://www1.eere.energy.gov/wind/pdfs/2012_wind_technologies_market_report.pdf.

[93] Philip Brown, "U.S. Renewable Electricity: How Does the Production Tax Credit (PTC) Impact Wind Markets?" *Congressional Research Service*, June 20, 2012.

[94] Eric Lantz, et al., "Implications of a PTC Extension on U.S. Wind Deployment," *National Renewable Energy Laboratory*, April, 2014, NREL/TP-6A20-61663. http://www.nrel.gov/docs/fy14osti/61663.pdf.

[95] "Production Tax Credit for Renewable Energy," *Union of Concerned Scientists* web site. http://www.ucsusa.org/clean_energy/smart-energy-solutions/increase-renewables/production-tax-credit-for.html.

[96] D. Rastler, "Electricity Energy Storage Technology Options," *Electric Power Research Institute*, December, 23, 2010. http://www.epri.com/abstracts/pages/productabstract.aspx?ProductID=000000000001020676.

[97] Greg Albright, Jake Edie, and Said Al-Hallaj, "A Comparison of Lead Acid to Lithium-ion in Stationary Storage Applications," *AlerternateEnergyMag.com*. http://altenergymag.com/emagazine/2012/04/a-comparison-of-lead-acid-to-lithium-ion-in-stationary-storage-applications/1884.

CITATIONS

[98] "Is Lithium-ion the Ideal Battery?" *Battery University web site.* http://batteryuniversity.com/learn/article/is_lithium_ion_the_ideal_battery.

[99] Eric Wesoff, "Isentropic's Pumped Heat System Stores Energy at Grid Scale," *GreenTech Media web site*, July 9, 2013. http://www.greentechmedia.com/articles/read/Isentropics-Pumped-Heat-System-Stores-Energy-at-Grid-Scale.

[100] Z. Ye, R. Walling, N. Miller, P. Du and K. Nelson, "Facility Microgrids," *National Renewable Energy Laboratory*, May 2005. http://www.nrel.gov/docs/fy05osti/38019.pdf.

[101] John Bradley, Assistant Vice President, New York University, Interview with author, April 10, 2013.

[102] Occupational Safety & Health Administration—Regulations (Stards—29 CFR)," *U.S. Department of Labor*, 2002. https://www.osha.gov/pls/oshaweb/owadisp.show_document?p_id=9726&p_table=standards.

[103] "Animals in Disasters: The Four Phases of Emergency Management," *Federal Emergency Management Agency*, 2010. http://training.fema.gov/emiweb/downloads/is10_unit3.doc.

[104] "National Incident Management System Training Program," *U.S. Department of Homeland Security*, 2011. http://www.fema.gov/pdf/emergency/nims/nims_training_program.pdf.

[105] "Emergency Operations Plan: Version 3.0," *Cornell University*, 2014. http://emergency.cornell.edu/files/2014/02/EOP_with_ESF_Annexes_v3_02FEB14-yda7ox.pdf.

[106] Lanny Joyce, Director of Utilities and Energy Management, Cornell University, Interview with author, conducted June 23, 2013.

[107] Brian Dosa, Director of Public, Fort Hood, *Interview with Author*, conducted on July 15, 2013.

Index

A
American Society of Civil Engineers 11, 132
ASHRAE 122, 193

C
California 10
California Energy Commission 10
California Hospital Association 184
Caroline, New York 49, 80
climate change 10, 15, 28
Commercial Building Energy Consumption Survey 136
community xi
Community Energy Planning Cycle 54
Community Energy Strategic Planning Academy 53
Cornell University 46, 81, 188, 192

D
Department of Defense 40
deregulation 13, 21

E
educational campuses xi, 45
Energy Independence and Security Act 118
Energy Independence and Security Act of 2007 115
energy resilience xi, 3, 27
ENERGY STAR xiii
ENERGY STAR Portfolio Manager 110

Executive Order 13514 118

F
Federal Emergency Management Agency 188
Federal Energy Management Program 41
Fort Bliss 42, 64
Fort Carson 42, 128
Fort Hood 198

G
Gundersen National Health System 44
Gunderson Health System 34

H
Hazard Mitigation Grant Program 48
healthcare campuses xi, 43
Hospital Energy Alliance 44
hospitals 43
Hurricane Sandy 19

I
Incident Command System 188
International Energy Agency 159
island mode 34, 171

L
Langone Medical Center 43
Lawrence Berkley National Laboratory 118, 161
LEED xiii, 32, 118, 128, 193
Local Law 87 118

M
military bases xi, 40
municipal governments xi

N
National Renewable Energy Laboratory 41, 129, 143, 149, 171
net metering 163
Net-zero Energy Communities 143
Net Zero Energy Installation 41
Net-Zero Military Installations 60
New York University 176
NYU 46
NYU Langone Medical Center 30

O
Occupational Safety and Health Act 180

P
peak demand 166
population growth 5

Princeton HealthCare System 7
Production Tax Credit 164

R
renewable portfolio standards 164
residential communities 48
Rockport, Missouri 49

S
Sandia National Laboratory 129
spectrum of energy resilience 31

U
University Medical Center of Princeton 45
University Medical Center of Princeton at Plainsboro 7
utility restructuring 13

DATE DUE